An Illustrated
History of Fa

GW01459122

Turning the wastes at hay making time.

An Illustrated History of

FARMING

by PETER WILKES

SPURBOOKS LIMITED

Published by Spurbooks Limited
6 Parade Court, Bourne End, Buckinghamshire

B 027629

ISBN 0 904978 84 2

Designed and produced by
Mechanick Exercises, London

Typesetting by Inforum, Portsmouth

Printed in Great Britain by
Tonbridge Printers, Tonbridge, Kent

Contents

Illustrations

Introduction

On that summer day in 1950, when steam enthusiast Arthur Napper accepted a challenge to race his 1902 Marshall traction engine, 'Old Timer' against 'Lady Grove' the 1918 Aveling owned by Miles Chetwynd-Stapleton, friend and fellow enthusiast, he little realised that he was setting a pattern which would help change our recreational habits.

For such was the public response to the event that an idea formed in Arthur Napper's mind — an idea which was to lead to the Traction Engine Rally which is now established in every part of the country.

But to re-kindle the age of steam, the period when traction engines proved their ability at a variety of tasks, particularly in the countryside, it was necessary to harness them to the machinery they had driven during their working life. Hence the old threshing drums, saw benches, chaff cutters and other reminders of an agricultural past found a new lease of life as they demonstrated skills and methods of working which at one time appeared to have been lost for ever.

Preservation however, in relation to agriculture, had gone further than just saving the implements associated with the traction engine. For as change inevitably made its way into the rural scene, far sighted people, recognising that a way of life was disappearing, began to rescue implements which were in danger of disappearing into oblivion, ensuring that what was once a way of life would not be lost for ever.

It was the Traction Engine Rally which brought a general awareness of the past and as public interest grew, so those concerned with our agricultural past added to the appeal of the rally by showing examples of agricultural implements of yesteryear.

Sickles and sythes, once the general tool of the harvest field, not only brought back memories of the days when both men and women toiled in the fields, but showed a new generation a method of farming which many believed impossible. There were flails, the universal tool of the threshing, showing the time-honoured system which was in use until the appearance of the threshing drum. The fiddle and broadcast drills that men used when they walked the fields sowing their seeds brought a surge of interest, showing a new awareness of rural life as it used to be.

It is not only the farming past that excites interest. Today more than ever before, the barriers between town and country are disappearing, as

A 1913 Ruston Proctor General Purpose traction engine, No. 46596 of the type used for varied work in agriculture.

the way farming operates begins to appeal to those to whom in the past it was always a mystery. The various County Shows are no longer attracting only those concerned with agriculture. Today they become events that bring the crowds flocking out of town and city to enjoy the general pleasures, as well as to see at first hand the way in which mechanisation has changed the face of agriculture.

Part of the appeal of the agricultural bygone lies in the contrast that exists between those early farmers and the men who get their living from the land in this modern age. The combine, the universal tool of the harvest field, is one example of the way progress has made its way into the agricultural scene. To buy one would cost between £15,000 and £32,000, although only very large farms could justify the latter price.

In general it is the smaller combine that is found on most of our farms; combines which, in a working day, will devour some 20-30 acres of corn as opposed to the 16 acres that a binder with two horse teams would get through on a good day. The combine harvester delivers the grain threshed and ready for drying. In contrast the sheeves from the binder had to be 'stooked' to take advantage of the drying effects of the summer weather. Then they would be stacked before threshing began, often with an outside contractor bringing in his traction engine and threshing drum.

Harvesting is only one example of the way mechanisation has become the motive force in modern farming, seeing the old time farm labourer give way to a class of technician able to operate and service the many complex implements that are now the lot of the farming community.

It is the aim of this book to take the reader through the agricultural revolution that has seen the time-honoured methods gradually give way in the face of progress. It has been written, not for the agricultural expert, but for the interested lay reader, someone who finds pleasure in our rural heritage, tracing the way the agricultural scene has changed, bringing efficiency into the farming scene on a scale that at one time would have seemed impossible.

I·The Harvest

Today the combine harvester is the universal machine in the cornfields, not only 'reaping' but threshing and cleaning the grain so it is delivered to the farmer ready for drying and storage.

The picture before mechanisation came to agriculture was vastly different. Then the taking of the harvest was a period of toil. Men and women used the sickle to cut the crop. During the winter months, when the weather made outside work impossible, the crop was laid out in the barn and the grain separated from the ears by means of the flail. Yet although the tools may appear primitive to our eyes, even the humble sickle was the product of mathematical calculations and subject to many patents as manufacturers sought to ease the lot of the harvest worker. In fact Henry Stephens, the noted 19th-century agricultural writer, when describing the final shape of the sickle, spoke of it as having a blade with 'the curve of least exertion.' This was the design purposely evolved so that not only was the sickle light in weight but of a shape that demanded the minimum of effort. It had to be a tool which the women could also handle through the long hot days in the harvest field, without fatigue interrupting the rhythm of reaping.

Its evolution went further than just shape, for the sickle developed into two distinct types. The Ancient Egyptians are reputed to have taken the jaw bones of animals for their sickles, sharpening the teeth into an efficient cutting edge. This was a trend that was to continue, for in the days of Victorian farming, Stephens enthuses over the serrated or 'toothed' sickle.

While iron continued to be the metal for the blade from the period of the Middle Ages, in many cases it became the practice for the cutting edge to be formed from steel, with serrations or teeth knocked into it with a hammer and chisel. Using this as a harvest tool, the reaper would clutch the corn with his left hand while using a sawing action with the sickle to cut the crop.

Reaping was a job that had to be carried out with forethought and planning. Every moment was precious, as men toiled to get the corn in before the weather turned against them. In fact, as Stephens describes when he writes of the bandwin method of harvesting, organisation of labour was very much to the fore. This system used seven harvesters working as a

A modern combine. It is shown here emptying its tank of combined grain into the trailer which will take it to the grain store.

team. Three reapers worked on two ridges or side by side widths of the crop, with one bandster serving the six. While Stephens refers to the bandster's work (the tying and carrying of the sheeves into the familiar stookes) as too hard for women, there was a definite place for women in the Victorian harvest field. In fact the ideal deployment of labour was shown as two women reapers working at the sides of each ridge, with a man in the centre who also prepared and laid out the bands which were used to tie the corn into sheeves. In practice, the bands were laid out, and each reaper put his corn on them until there were sufficient to make into a sheaf of accepted size. Working like this one man could handle the output of the six reapers and the team as a whole could get through two acres of corn per day.

The other form of sickle was the smooth edged type that really came

Close up of the action of the Thirsk-built reaper.

into its own if the corn had been flattened by wind or rain. Then this sickle would cut through the corn at a stroke, not requiring the reaper to grasp the crop as he had to when using the serrated edged tool.

Generally the smooth edged sickle was the quicker of the two, but it needed constant sharpening whereas the toothed type could be used during the harvest without such need; a factor that greatly added to the value when reaping began.

The scythe of course, was another implement of the harvest field, but there were many who saw it as inferior to the sickle, which was said to preserve the grain better.

The tool that was first seen as a replacement for the sickle was the 'Hainault' scythe that originated in Flanders during the Middle Ages. Travellers from Europe returned with stories of how it could cut up to one

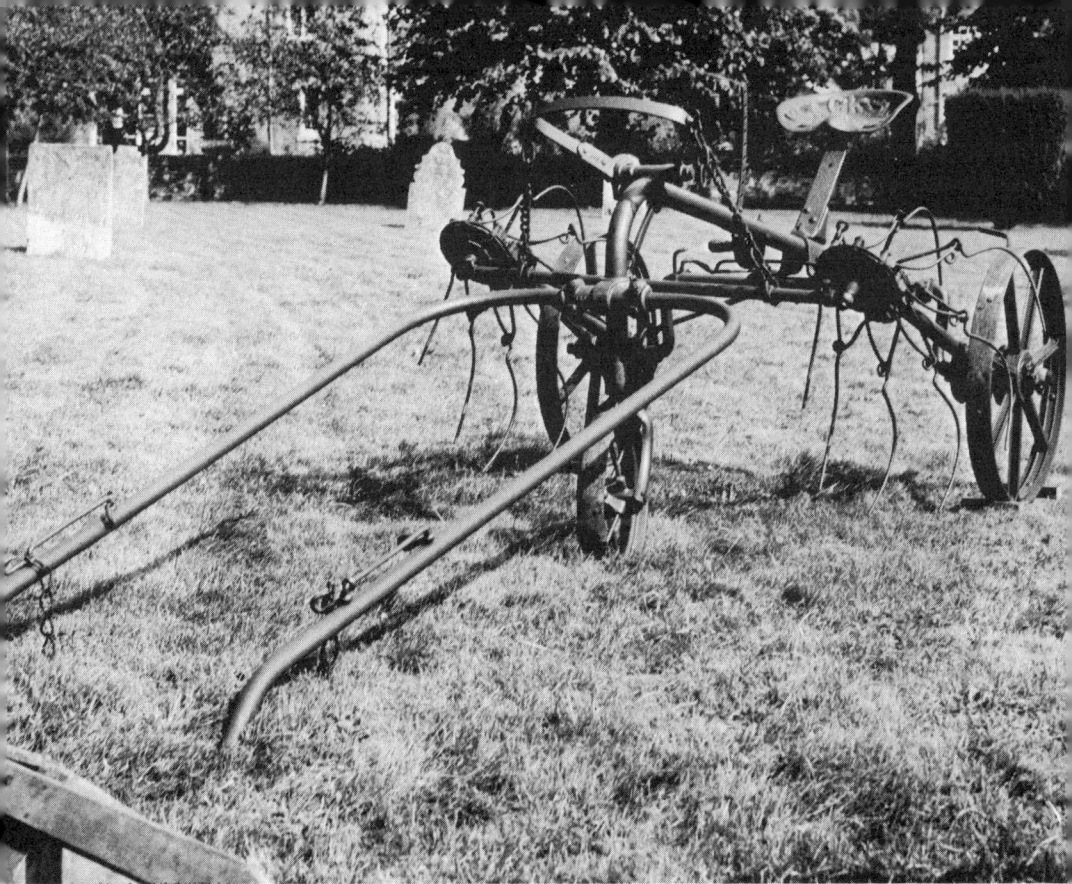

A horse-drawn hay tedder, the implement of the harvest field in the age of the horse, used for turning the hay.

third more corn than the sickle, and many farmers in an attempt to get their men on to this tool gave them 'Hainault' scythes as presents.

Although they were smaller than the native English scythes, they were awkward to use and most harvest workers reverted to the sickle. However, in many areas during the 19th century a form of scythe did replace the sickle in the corn fields. This was the cradle scythe, a vastly different tool from the Flemmish import that had been tried and rejected. The blade could be anything up to 48″ in length and an important feature of the reaping scythe was a wooden cradle, made possibly of three horizontal wooden spikes. This was fitted parallel to the cutting blade and ensured that the cut crop was laid out in a regular swathe across the field. Without the cradle, the corn would have fallen in all directions causing the gatherers to spend far more time forming the sheaves. Again, working

17

A selection of corn dollies.

An early combine harvester — Massey Harris 21 c. 1943.

with the scythe was an organised matter, with Henry Stephens writing of how a unit of ten undertook the task. This was made up of three scythemen to do the actual reaping, three women to gather the corn, three more men to tie and stook it, and one man raking. Another alternative was for two scythemen, two gatherers, two bandsters and a woman to rake the stubble. The system used depended on the size of the farm. On the larger establishments the labour force would be organised into a number of these units, while a small farm would use only one unit of ten to carry out the work.

In the days when men and women used the scythe and sickle in the cornfields, harvest time was more than just work. It was the highlight of the farming year, and a period more than any other when tradition truly came into its own.

For as long as man can trace back into history, tradition, in particular the 'corn dolly', has been associated with the rituals of harvest, although no one has been able to discover its origins. The custom of offering thanks for a bountiful harvest, and the hopes of a successful yield the following year, varied from civilisation to civilisation. Yet whatever the form, the motive was the same.

Either Ceres the Corn Mother, or Demeter the Goddess of Fertility, was thanked as the awakener of life, and Cornucopia, the Horn of Plenty, was their symbol of the future. Such is the age of these straw figures that today it is only possible to surmise at the meaning of the word 'dolly'. But since the harvest was concerned with the fertility of crops, it is inevitable that any figure made to be worshipped would be female. The corn dolly was the symbol of joy and relief after the harvest had been safely gathered in. Also after the seed had been sown for the next year's crop, a traditional straw figure was carried ceremoniously round the fields in the hope that it would bring another bountiful harvest. But it was in accompanying the 'last load' that the dolly had its true place in rural life. Then it stood as a symbol of the merrymaking that followed the release from the fears of bad weather stopping the harvesting and ruining the crops. In some parts of the world the corn dolly was a figure of worship, but in Britain it was a focal point of decoration, an example of how our ancestors, despite the hardships of a farming life, proved their ability to celebrate. With the passing of the years, the harvest celebration became as joyous an occasion as Christmas.

With the 'last load', work could be forgotten for a short time. The farm waggon was gaily decorated, the horses fitted with shining harness, and both men and women donned their best clothes. The lead horse would be ridden by the prettiest girl in the village and the procession would make its way, often going to the church for the harvest service before going on to the farmhouse where a feast would be provided by the farmer for his workers and their families.

The role of the corn dolly in the celebrations varied from county to county. The figures were usually made by the older people who were no longer fit enough to take a hand in the actual harvesting. When finished they would provide decorations for the church, the farmhouse and the procession from the field. Often a dolly was ceremoniously carried into the farmhouse, given pride of place during the harvest feast and then put away as a symbol of good luck for the following year.

In Devon a singular feature was the harvest cry. When the last corn had been cut, the reapers assembled round a corn dolly and joined in a ceremony which finished with them all shouting, *"The end,"* or something similar, three times. By this means the news that the harvest had been successfully gathered in was spread round the village. In other areas the end of the harvest was signalled by a young man taking a dolly made from the last corn to be cut and racing to the farmhouse with it. He had to get the straw figure into the house without being seen. Failure meant a soaking, for the women folk were waiting, all armed with buckets of cold water! The form of the dolly varied, although Mother Earth and the Horn of Plenty were favourite symbols common to all regions. In Cambridgeshire, where the designs follow early traditions the 'last load' made its triumphant procession from the fields accompanied by men ringing hand bells, hence the *'Cambridgeshire Bell'* corn dolly.

A reminder of the age of the horse in the harvest field comes with the Suffolk dolly. This was the county famed for its powerful draught horse, the Suffolk Punch, so their corn dolly took the form of a horseshoe and whip. However traditional, straw figures are not always easy to explain.

The threshing drum before the advent of the combine harvester.

Norfolk, has its 'Corn Lantern' whose origins are still open to debate. Other designs that add to the fascination of this aspect of our rural heritage include the Vale of Pickering 'Chalise', the 'Ive Girl' from Kent, the Northamptonshire 'Horns' and the Staffordshire 'Knot'.

Alas, with the coming of the combine harvester the corn dolly and the traditions it served became a forgotten part of the countryside. There is no place for sentiment in the business of mechanised farming. Today this aspect of the harvest scene has been preserved by enthusiasts who value rural life as it used to be. At country meetings of every kind, from the humble village fete to the grandeur that is now the County Show, there are those who, having dedicated themselves to learning what could so easily have been a lost art, show their skill for the benefit of us all.

However, the corn dolly was still a part of rural life when mechanisation first made inroads into the farming scene with the introduction of the reaper. The reaper however, far from being a Victorian idea, was envisaged in the 18th century. As early as 1783 the Royal Society of Arts offered a prize for a machine whereby wheat, oats, barley or beans could be reaped more expediently and cheaply. Yet although the need was recognised, man's engineering capabilities were not yet up to the standard demanded. In fact for the next hundred years it was the old sickle and scythe which served as the standard tools in the harvest field, although experiments with reapers were made from 1800 onwards. The first designers of mechanical reapers used a circular iron plate, to which either scythe blades had been attached or the plate itself sharpened and notched to give a circular saw action. One of the main drawbacks was that the corn was thrown about with no provision for laying it, something that James Smith of Deanston thought he had overcome in 1811.

James Smith used the idea of a continuous revolving knife, turning horizontally just above the ground, with a canvas drum above the cutter, against which the corn would fall and be thrown on to the ground in regular rows. In action however, the variation in the speed of the horses operating the reaper affected the rotary speed of the drum and the corn was thrown about the field.

In fact it was only when the principle of the reciprocating shears was applied to the reaper that success was achieved. The man who brought out the first successful reaper for Britain was the Rev. Patrick Bell, a Scottish minister. Bell's reaper was pushed by two horses, one either side of a pole with a handle at the end for the man in control.

The cutting action was achieved by a row of triangular blades which

A sail reaper being pulled by a tractor. In its time this would have been pulled by horses.

were sharpened on both edges and fixed a few inches above the ground. Reciprocating blades, cutting both left and right, were fitted under and between the stationary blades. It embodied many of the principles which were common to reapers of a later age, with a revolving, six bladed 'sail' throwing the corn on to a travelling canvas which deposited it in rows on the ground.

Although Bell's reaper did the job it was intended for, only a few were built and used in the harvest field. Henry Stephens, writing about Bell's reaper in his *'Book Of The Farm'*, saw the great disadvantage in having horses working at the back of the reaper which were in general difficult to control. In the field of mechanical reaping, Stephens recommended two American imports made by McCormick and Hussey that had caused such a stir when first shown at the Great Exhibition of 1851. McCormick relied on an oscillating action to saw into the crop, while Hussey had a series of triangular blades which moved between two lines of projecting guards.

Both methods were effective, but Stephens, making a choice between the two, saw the McCormick reaper as superior. His decision made sense in practical farming terms, for whereas the McCormick implement laid the cut corn out of the way, enabling it to work continuously without stopping, the Hussey reaper laid the swathe behind it, so that it had to be cleared before the reaper could go on to the next cut. Both reapers were made under licence, but the Hussey one was the cheaper of the two, so many farmers put up with the inconvenience of having to clear the cut crop after it in return for a lower financial outlay.

The Great Exhibition of 1851 was the turning point in bringing mechanisation to the harvest field. Within two years of that date Hussey alone was claiming sales in Britain of 1,500 reapers. Even so the sickle and scythe continued to dominate, but now they had been shown the way, other manufacturers entered into this aspect of implement building. In fact at trials held in 1869 by the Royal Agricultural Society in Manchester, eighty-four different types of reaping machine took part, and within two years of that date it was said that a quarter of all our corn harvest was being reaped mechanically.

An early reaper, one of the first to succeed the scythe and sickle as the standard mowing tool — built at Thirsk, Yorkshire.

A Massey Harris combine harvester, 1958.

By this time a further advance in the form of the 'sail reaper' had appeared in the harvest fields. It took mechanisation one stage further by using four revolving sails to scrape the cut corn off at intervals, leaving it in regular bunches on the ground. Of course, it still had to be gathered by hand and tied into sheaves, so the next evolution in the harvest field was to be the 'binder', the machine that could do this task automatically. It was America which produced the first practical binder. Two Illinois farming brothers, W. and C. Marsh patented a 'harvester' in 1858. However, provision was only made to carry the cut corn over a canvas belt to men who still had the job of tying the sheaves by hand.

The first mechanical binding mechanism, again from America, came in the same year as the Marsh offering. It was effective but used twine, a commodity that in those days was expensive, so for general farming needs it

25

was dismissed. It was not until 1871 when Walter A. Wood, another American, produced a binding mechanism that secured the sheaves with wire. Its use was short lived. By 1880 the cost of twine had slumped and farmers found wire damaging to both their machinery and animals. This was the signal for the re-emergence of the 1858 sheaf-tying mechanism and the move towards its use in the harvest fields.

In fact the binder, as it came to the farmer, was a combination of this tying mechanism, the McCormick cutter, and the Marsh system of canvas belt to transport the cut corn along.

This became the universal implement of the harvest field until the arrival of the combine harvester. The frame of the binder was made almost entirely of steel, with a steel platform riveted to it, and a 'grain' wheel supporting it at the free end. In order to prevent this wheel from running over part of the crop, a wedge-shaped divider was fitted in front of it, which not only prevented the standing corn from being damaged, but also raised fallen shoots as they were cut more readily. Special dividers could be fitted over the 'grain' wheel for use where the crop was flattened through the effects of wind or rain.

The cutter bar was positioned in front of the platform and provided with a lever to regulate its height. The 'reel', the six bladed beater which bent the corn towards the knife, could be adjusted vertically and horizontally and placed in whatever position was best suited to the height of the crop and the position in which it was leaning.

As the corn fell onto the platform it was received by a bottom canvas and carried by the two elevating canvases to the deck of the binder. There packers and butters compacted the corn into a sheaf and carried it to the knotter. A curved and threaded needle arm carried the twine round the sheaf, and left it in the grasp of the knotter, which tied and cut the twine before the sheaf was thrown off the deck.

While the binder was an advance on anything that had previously been known in the harvest field, it still left the threshing, the removal of the grain from the ears, to be done by hand.

In the early days of farming this was a back breaking job done on a specially prepared part of the barn floor with the flail. The flail was the universal threshing tool from the Middle Ages through to the end of the 19th century. Before the use of the flail, which was nothing more than a handstaff some four to five feet long with a beater and another circular piece of wood joined to it by a leather thong, the grain had been knocked out with a simple stick.

26

Wire tying baler used with the combine.

Threshing was a job done in the barn when winter weather made other work impossible, and while it was a laborious job, it was also one needing skill. The thresher had to swing the handle over his shoulder and bring the swingel down so that it struck the corn just below the ears. Hitting it in that position ensured that the grain was shaken out without being damaged in any way. During the periods of threshing the doors in each wall were opened so there was a constant current of air to carry away some of the dust created. After the threshing the grain was sieved and then heaped on the floor ready to be cleaned. This cleaning was done by throwing the grain in to the air, often into an artificial draught. Also known as winnowing, the lighter chaff and smaller grains would be blown further away, while the perfect heavy grain fell into a separate heap.

The first machine for winnowing was built following a Dutch pattern and consisted of rotary sieves and canvas sails to create the breeze. This

International Deering self-tying binder built in America. Used between the turn of the century and the late 1940s.

was replaced by an internal rotary fan and winnowing machines in various sizes became a familiar feature on the farm.

However, it was a replacement for the flail that agriculture badly needed. Not that men were without ideas. As early as 1732, Michael Menzies, a Scotsman, produced his version of a threshing machine, and proved its ability by re-threshing corn that had already been flailed, and recovering grain from it. However the basis was still the flail, which was made to strike the corn as the machine revolved, and in use there was a constant flail breakage rate that, in practical farming, made it impractical to use.

Another concept when considering mechanical threshing was the danger of bruised and damaged grain, and, many otherwise efficient designs suffered from this drawback. The breakthrough came in 1786 when Andrew Meikle of East Lothian, devised a system whereby the grain was separated from the ears by a revolving drum — the 'threshing drum'. Meikle had given the farmer the system that had eluded others and a design that was to serve through to our own age, for the threshing mechanism of the combine harvester does not differ greatly from the old stationary threshers. What Meikle did was to provide four longitudinal

28

A modern combine harvester capable of cutting and threshing up to 50 acres per day.

beater bars on the outside of the drum, which in his case was turned by water. The arrangement allowed a space of about one inch between the beater bars and the covering of the drum. Corn was put in, ears first, between two rollers and passed through the drum so the ears were struck on the upward turn of the beaters, knocking out the grain, before the sheaf had cleared the rollers.

Later improvements saw shakers to further separate the grain as it fell through into riddles, and while the number of beater bars had to be increased when steam power was harnessed for threshing, it is a tribute to the ability and ingenuity of Meikle that his basic idea has survived in our modern farming implements.

Meikle's machines, powered by water, horses or even the wind, were permanent installations, and it was not until the early years of the 19th century that smaller portable threshing drums came into use.

The influx of the threshing drum was a cause of the agricultural workers unrest that led to so much violence in 1830, when farm workers, fearing that the new mechanical introductions would lower their own living standards, attacked and destroyed so many machines. But progress can never stand still, and after concerted action by farmers, the threshing

29

drum became as much a part of the agricultural scene as the reapers and binders.

Yet even as the threshing drum was attaining the perfection of its later years, on the other side of the Atlantic men were working on the idea of an implement that would do everything as the corn was being cut. For while we tend to think of the combine harvester as a recent innovation, it was in 1836 that the first one appeared in America. This was horse drawn, of course, and although it performed well, 'combining' by 1843 twenty-five acres a day, it was not until 1928 that the first one appeared in Britain.

Conditions between British and American farms were vastly different. As the 19th century came to a close and the combine was finding favour on the large American farms, in Britain there was adequate labour to work on the smaller holdings and in terms of expense the combine could hardly be justified. In fact the binder and threshing drum were still the main tools of harvest at the outbreak of the Second World War, with only a handful of farms using a combine.

Today of course, things are vastly different. It is the combine that predominates the harvest scene, working the corn at a speed previously thought impossible. In fact such is their efficiency that in an experimental scheme, over one hundred acres of cereals were combined in one day, using one of the latest of the 'giants'. Even so, on a farm geared to its cereal crop, one or more combines capable of taking in 45 acres of the harvest in a working day will be found.

While it appears that the combine is the ultimate in harvest machines, there has been a recent experiment in Sweden where the wheat is harvested as a whole with the sileage type of cutter blower, and then threshed indoors with the resultant straw being used to heat the corn dryer. So in the future there could be changes that will bring as drastic a revolution to the farming scene as occurred when it moved from the age of steam into the period of the internal combustion engine. Of course, there would be no use for machinery if there were not the crops to harvest, and here the farmer has to make his choice as to what he grows on his particular farm.

The cereals, wheat, barley and oats, have from the early days of farming, produced a large proportion of our food and that of our animals. Wheat is the crop for our bread and biscuits, barley is used for brewing and distilling, and a small proportion of oats is also needed for human consumption. Public preference must be taken into account when crops are being considered and to this end Canadian and American wheats give a flour resulting in the light loaves that are the favourite with housewives.

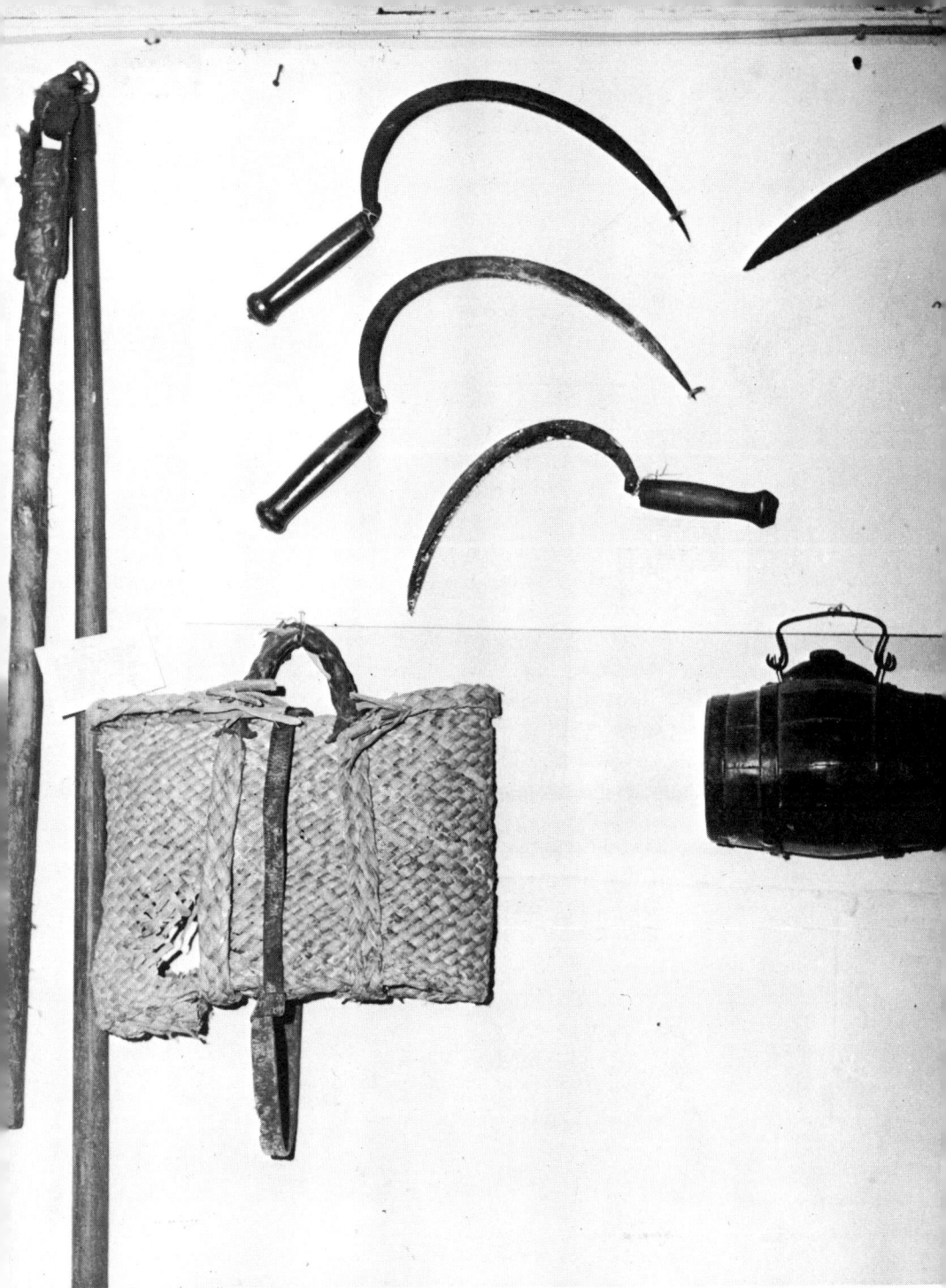

A reminder of the early days of harvesting; showing the flail, serrated-edged sickles, a harvest basket and a harvest beer barrel.

Other types that are grown in this country, can give a higher yield but are not ideal for bread making. This is where the research chemists come into their own, developing wheats that will give the required yield and at the same time the quality that the buyer demands.

Barley in turn, must have a strong uniform germinating power if it is to prove suitable for malting, the use that brings the highest financial return. So a good knowledge of his soil is essential to the farmer when he is deciding which variety to plant, otherwise he risks the crop finishing up as animal feed, with a resultant large drop in profit.

The choice between wheat and barley is mainly decided by the type of land and the crop rotation that is being followed to get the maximum return from the soil. Again, when it comes to planting the farmer has the choice of a spring or winter sown wheat. In the main, winter wheat sown between October and December is used, for the yield can be up to 25% higher than spring sown wheat. In contrast, most barley is put in the ground between February and April. Oats, which are not grown to the same extent as wheat or barley, can be either spring or winter sown. In practice in the South of England both types are used, but as one progresses north it is the spring sown varieties that are planted.

In the past, the cereal farmer was concerned with only wheat, barley or oats. However, in recent years another crop has found its way on to the farming scene. Rape, the yellow flowering crop, is grown for its vegetable oil, and is increasing in acreage every year. The reason is the world shortage of vegetable oil due to developing nations using it themselves instead of exporting what they grow. Rape is sown in early September and is cut in July, but harvesting differs from other cereals. Rape is first mown similar to hay and laid out in swathes on the stubble for 7-14 days. The reason for this advance cutting and weathering is that it is a crop which ripens unevenly. So early cutting and a period exposed to the sun gives the benefit of even ripening, and makes it ready for the combine harvester which comes into the field to deal with the cut crop in the same way as it does with other cereals.

2·Ploughing

Whatever the choice of crop, success in farming depends on the soil receiving adequate preparation prior to sowing. The gardener, when he is getting his land ready, turns the soil over with a spade, exposing it to the breaking down action of the atmosphere and winter frosts.

The farmer's needs are exactly the same, but in his case, with a large expanse of land, he uses the plough rather than the spade, but in those early days when men first began to wrest a living from the land, the plough was unknown, so they would turn to the materials at hand, working with nothing more complicated than a 'digging stick', a piece of wood possibly cut from a tree with the end sharpened into a point. Later a projection was fitted at the bottom so the action of the foot could assist as it was pressed into the soil. The digging stick, in turn gave way to a primitive form of hoe. Although little more than a forked branch with a side cut short and pointed to scratch into the surface, it made the task easier as men hauled it along behind them.

Even when the Ancient Egyptians came up with a form of plough fitted with a rectangular iron share, it was still only a cutting tool that scratched rather than turned the land over. In fact, for complete cultivation, it was necessary to plough and cross-plough the fields. The Celtic plough, in the pre-Roman age, was itself hardly more than a digging stick, scratching the surface rather than turning it over in the way we understand today. The Romans however, when they arrived in England and established their farms, had ploughs in advance of anything that had been previously used in the country, being fitted with a form of coulter and wings that acted as a form of mouldboard.

Part of the confusion which arises when people study the various types of plough to be seen at the many agricultural meetings and shows where implements of yesteryear work again, is due to the number of terms that are associated with a particular implement. Although many of the names given in the past varied from area to area, the modern plough has a British Standard Glossary of Terms, which gives the names and definitions of the principal parts.

The beam, for example, is defined as that part to which the power is applied and to which the other parts may be attached, either directly or indirectly. Today the facility for using various parts according to the type

A plough from the age of the horse with a skim coulter for turning in surface growth to the main furrow.

of work being done is found on most ploughs, and provision is made for individual adjustment, for, as many ploughmen will relate, good ploughing only results if the man knows his plough. This means that he will adjust its settings, and calculate everything to the smallest degree; something that can only come about through practice and experience.

The plough-share is the cutting part of the plough, the piece that thrusts into the soil. The coulter is the part of the plough that makes a vertical cut in the soil ahead of the share. Coulters come in three types. The 'knife' coulter fastens to the beam by a clamp which allows adjustments to suit the type of work being done. The 'disc' coulter is used where the knife coulter would cause difficulty by becoming blocked through weeds or straw getting wrapped around it. The disc coulter, a free spinning circular disc with a cutting edge, does not encounter such trouble. An even better form of coulter for use in ploughing through straw is the wavy-edge disc coulter. The third form of coulter found on ploughs is the 'skim' coulter, which is used in addition to the knife or disc to assist in the burial of manure, rubbish or crop remains. In effect the skim coulter consists of a miniature plough fixed to the beam in front of the knife or just behind the disc, so that it throws the straw that it has pared off the land into the bottom of the furrow.

The 'mouldboard' or 'breast' continues behind the share, appearing in fact, to be one with it. It is the mouldboard that moves the soil sideways and turns it over, and plough types are determined by the kind of mouldboard fitted. The four principal ploughs in order of decreasing length and increasing depth of the mouldboard are, the 'ley', 'general purpose', 'semi-digger' and 'digger'. The 'ley' and 'general purpose' have a mouldboard with a very gradual curve and a slightly convex surface. The 'digger plough' has a mouldboard that is short and abrupt with a concave surface, while the 'semi-digger' falls in between the 'general purpose' and the 'digger'.

It is the differences in length and curvature which determine the degree

Demonstration of horse ploughing.

The anti balance plough.

of pulverisation of the soil. The 'ley' and 'general purpose' ploughs turn an unbroken furrow and press the slices firmly together. With the other two ploughs the soil rises up the mouldboard and falls in a broken and pulverised condition.

An implement such as this however, was not even dreamed about in the early days of agriculture and man had to work from his primitive digging stick, evolving gradually until the basic pattern of the plough emerged. An interesting reflection of those lost years of farming is that it was a requirement by law in early England that each ploughman made his own implement and no one was allowed to plough until they could do this.

It was in Holland that the first attempts were made to bring the plough forward in design and it is possible that it was the Dutch influence which brought out what was known as the 'Rotherham' plough, introduced about 1730 by Joseph Foljambe at Rotherham, and described as "the most perfect implement then in use."

It had no wheels, being a 'swing' plough, and it was made of wood but had an iron share and coulter with a mouldboard covered with iron plate. This plough was used extensively in the North of England and had the distinction of being the first factory-made plough.

The same plough was introduced into Scotland where it was called the Dutch plough and although it found favour, even an implement as efficient as this could not alter the attitude of many farmers who followed old traditions, sticking to a form of plough that had been used by generations, even if it meant opening the land up with a separate implement before the plough could be taken onto the land.

It was however a Scotsman, James Small, who first applied mathematical considerations to the design of a mouldboard. In the past these had been made with a haphazardous twist according to the beliefs of the maker. Small saw the need for a single, universal type that would not only make ploughing more efficient but demand less animals to pull the implement. His offering to the farming community in the second half of the 18th century, was described as "reducing the expense of cultivation and being suitable for use in any soil". James Small's work paid off. The plough found favour in England as well as in Scotland and a factory to make them was opened in Ireland. The plough reigned supreme for a number of years, beating off many competitors, although there were other outstanding ploughs of that era, including one made by an Essex man, John

A modern multi turn over plough behind a crawler tractor on heavy, wet land.

A horse drawn plough and drill used for such crops as beans where the land needed to be ploughed to a depth of 3″—4″ for sowing.

Brand. His iron 'swing' plough intended to be drawn by two horses and later called the 'Suffolk' plough was made completely from iron and, again, was one of the ploughs that found acclaim among practical farmers.

There were others working towards the perfect implement for ploughing the fields, and one who did much to advance the design was Robert Ransome. In 1785 he obtained a patent for tempering cast iron plough shares, something that saw his business advance through demand from farmers and merchants, until he was in large scale foundry production. In 1803 he had further success when he obtained a patent for case hardening shares. Ransome chilled the under surface of the share more rapidly than the upper, making it harder so that wear was slower on the lower surface.

A mowing machine from the age of the horse, adapted possibly for cutting a pea crop.

The result was a self-sharpening share, for the quicker wearing upper edge resulted in the share retaining its sharpness.

But possibly the greatest advance that Ransome brought about in the plough came in 1808 when he patented yet another idea. This time it was for a plough that could be dismantled and also have new parts bolted to it. It was designed to appeal to the smaller farmer in particular. In the past it had been a case of buying a different implement for each job, but now with Ransome's offering they had one basic plough that could do a number of jobs in the fields, with tines available for sub-soiling and hoeing, and even seed boxes could be fitted.

Manufacturers however, had to do more than just make ploughs. An essential for success was a sales campaign where the merits of the plough could be demonstrated to farmers. To this end, Hornsby of Grantham, was but one manufacturer who saw ploughing matches as the perfect medium for advertising.

Ploughing became a demanding and skilled task, and the men responsible for this aspect of the farming scene were devoted to their work and proud of their ability behind the plough. So, in the close-knit community of the village, inevitably there were arguments and discussions as men compared their work with that of a neighbour. This led to the ploughing match that still has a place in rural life. Its origins date back to the distant past with the first recorded one having taken place at Odiham, Hampshire, in 1784. Gradually they became a scene of festivity as the community flocked to see its favourite in competition with others.

While there may have been a carnival atmosphere for the spectators, for those engaged in the ploughing it was a deadly serious occasion. In fact it was the intensity of the occasion that did so much to improve the standard of ploughing in general. In the beginning ploughing matches were local affairs, but as road systems developed and transport became easier, so the classes began to include the 'Open to All England' contest. To reflect back to one particular 'Champion of England', the Grantham (Lincs.) ploughman, is to capture a lost way of life. He won his title for the last time in 1922 at the age of 71 years. At the peak of his brilliance Miles Hardy was sponsored by the local agricultural implement makers, Hornsby of Grantham, with a plough specially made for him by their craftsmen. Sponsorship, with firms seeking acclaim for their implements through men such as Miles Hardy, had to be worked for.

Like so many other countryfolk Miles Hardy automatically followed in his father's footsteps, himself a famed horse ploughman. Such was Miles' skill and dedication, that by the time he was sixteen he was entering his

A 1920 4hp ploughing engine designed for direct ploughing, built to counter the threat of the tractor.

A modern multi furrow turn over plough of the type in use today. The disc coulter can be seen at the front with a skim coulter also fitted. The plough share and mouldboard are immediately to the rear of the coulters.

first match, but with a plough very different from the one Hornsby made for him. His first one cost the princely sum of 5/-, but in its competition life won him over £200 in prizes.

The rivalry between the competitors was intense in such competitions and not always good natured. In a letter to the *'Grantham Journal'* in 1905, a writer not only challenged the right of Miles Hardy to his title, but also a Mr. B. Noble over his title as 'Champion of Rutland'. The two men whose honour was at stake acted in the only way possible, and issued a challenge that the newspaper was pleased to print. "If the gentleman who has tried to pull Mr Hardy and Mr Noble down a peg or two has got any pluck, Mr Hardy and Mr Noble will plough any two first prize winners at any competition, for five pounds a side. The four to plough together and be judged as any other class, the winner to take ten pounds whichever side he might be ploughing on." The challenge was not accepted and after one more attack by the anonymous writer, the editor decided the correspondence was closed, although he relented to let another ploughman have the last words. "If the critic can plough as well as he can talk, why did he lose five consecutive matches to Mr Hardy?" From the pen of someone who obviously knew his identity, it brought forth the general comment, "Why indeed."

There was one drawback with early ploughs and that was the fact that the mouldboard turned the soil in only one direction. In terms of practical ploughing it was necessary to plan the work, so that at the end of the furrow the plough is taken to a different part of the field for the return cut. To have turned the plough and come back immediately at the side of the furrow would have meant the mouldboard turning the soil in the wrong direction.

An early plough that could be used continuously up and down the field was the 'turn wrest' plough. This was made with two mouldboards, each brought into use alternatively. In practice, one mouldboard would turn the furrows to the right, with the other mouldboard turning them to the left on the return journey, leaving all furrows lying in the same direction at the end of the ploughing.

Naturally as men worked the land their ideas turned to the concept of the multiple plough, an implement that could turn over two or more furrows at one pull across the field.

As early as the 17th century a double furrow plough came into existence, but it was heavy and cumbersome and required a powerful animal team to pull it. In fact it was not until the 19th century that a double furrow plough came into use. As more and more acres were being put under the plough to provide for the needs of an expanding population, the double furrow plough was one way of increasing output yet at the same time keeping costs down. For not only was it more economical with labour, three horses harnessed to this form of plough could do the work of four horses with a single furrow implement.

Today of course, the multiple plough is the rule rather than the exception, with anything from two to five furrows capable of being worked at the same time. Popular in modern farming are the 'turn over' multiple ploughs that can be turned over on a round beam when the tractor gets to the end of the row and is ready to make the next crossing of the field. Swinging the other half of the plough into the ground gives a furrow cut that lies in the same direction as the previous one. Again this is not a new idea, for Victorian farmers also used them in small numbers, as indeed

42

A modern sub soil plough which would be pulled by a crawler tractor.

they did with a lighter form of the balance plough from the age of steam.

Tractor development saw the controls for the plough within reach of the driver, while the modern tractor has the plough matched by a system which gives hydraulic adjustments from the driving cab.

All this, however, was far from the minds of men when they turned to the plough as an answer to the serious land drainage problems associated with farming on wet soil.

In Roman times the problem had been recognised and much marsh land was reclaimed by the digging of open drains and ditches to carry away excess water. The covered drain, nothing more than a deep trench cut into the soil and covered with brushwood and stones before the top soil was replaced, was a common method of drainage, but one that had serious disadvantages. It could easily become blocked as soil swept through it, and the drain itself often collapsed. Because of this many early farmers dug deep furrows to take away the surface water as well as acting as land boundaries.

Inevitably the need for drainage brought into being a special drain plough. In its simplest form, as it appeared in the middle of the 17th century, it was a device intended to cut down into the soil on either side of the intended drain, with the work of removing the soil between the cuts being done by hand. An improvement came in the early part of the 18th century in the form of a draining plough capable of digging out a drain some 18" by 12". The actual drain that the farmer used for his trench depended on individual choice. Some used the old brushwood method, others used large flat stones laying them just off the bottom of the trench. Systems in fact that were in use until the appearance of the 'horseshoe' type of land drainage tile which came into extensive use towards the end of the 18th century.

The breakthrough in land drainage came with the advent of the 'mole' plough, introduced in the same period when the drainage tile came into common use. The mole plough, although it carried the name 'plough', was nothing like the plough used as an implement of cultivation. In fact it consisted of nothing more than a knife edged coulter to cut through the ground, and a 12" long, 3" diameter circular 'mole' that cut through the ground leaving a drainage tunnel behind it.

The honour of inventing this form of land drainage implement went to Adam Scott, an Essex farmer whose mole plough was shown to the Agricultural Committee of the Royal Society of Arts in 1796. The man who made the next real development was the famous steam plough maker, John Fowler.

When Fowler introduced his mole plough at the 1851 Royal Show, it was the first of his many contributions to agriculture which were to amaze onlookers. For not only did it 'mole' a hole in the ground, but also laid a row of wooden drain pipes, a great contribution to land drainage. This was done by threading the pipes together and attaching the wire rope that ran through them to the plough. As the 'mole' cut its hole, so the pipes were pulled along after it. At the end the rope was recovered and a line of drainage pipes had been installed with only a mere slit on the surface of the land to show where it had been worked.

A requirement of adequate land drainage, particularly on really wet ground, is the breaking up of the hard packed sub-soil to allow the surface water to make its way through to be picked up by the land drains. It is also a requirement of farming that the sub-soil is not mixed with the top layer of soil. Although a form of sub-soil plough had been in existence for some years, the credit for the first really successful sub-soiler must be given to James Smith, a Scottish farmer who produced his version in 1823. It was an unwieldy implement, having a length of 15′, but was capable of ploughing to a depth of 20″. In operation the points of the share and coulter cut into the sub-soil, which was crumbled by a wing fitted to the back of the share.

The only real drawback to Smith's offering was that it gave its most effective results when preceded by an ordinary plough, a fact that caused designers to look at the possibility of a combination of sub-soil plough and ordinary plough. The sub-soil plough that set the pattern for the rest of the 19th century consisted of a single sub-soiling tine fitted at the back of the beam. Adjustable in its working depth, it was curved in shape so that the sharpened edge could cut in and stir the sub-soil, after the ordinary plough share and coulter had cut their furrow.

Sub-soil ploughs needed considerable effort to haul them through the ground, and in the case of Smith's plough at least six horses would be used. Today, with the more modern sub-soil ploughs it is usual to find a crawler tractor being used, particularly if the normal plough is working to a depth of between 7″ and 10″, with the sub-soiler tine going an additional 6″ into the sub soil.

A modern approach to this land preparation is the use of a chisel tined plough that breaks up the consolidation of the ground without turning the land over. After soil has been left to the pulverising effect of the weather, to go over it with a conventional plough would see the top surface that has received the effect of nature being turned over and the bottom soil brought instead to the top. To go through with chisel tines ensures that the ground is worked but leaves the top soil as the seed bed. On certain types of soil there has even been a move to do away with ploughing completely. Weed growth can be controlled with chemicals and special seed drills are in existence which can deposit the seed direct. Success here, however, depends on the type of soil on the farm.

3·Preparing the Land and Sowing the Seed

Ploughing is only the first step in getting the land ready for the seed. After it has been left exposed to the weather, the furrow slices have to be broken up and the soil worked into a tilth. The implements used for this aspect of the farming scene are the harrow and cultivator, which also serve for destroying weeds and covering the seeds. There is no essential difference however, between these implements, but in general, the cultivator is used to do the heavy work and large clods, while the harrow will prepare the fine tilth necessary for the seeds. Today, there are implements, called 'cultivator-harrows' which can be used for either deep or shallow work, but for the early farmer implements such as this were mere dreams in the future. He had to content himself in the beginning with what nature provided. It is therefore possible that our farming ancestors used nothing more than a thorny branch cut from a tree to break down the land that had been worked with their digging tool. In fact a form of 'bush harrow' was used in some countries up to the end of the 19th century. It was the obvious implement to follow the use of a tree branch and was made in the form of a sledge with thorny branches fitted between cross pieces, and projecting out at the end to give the harrow effect when pulled over the soil. The ease with which they could be made led to home construction on the farm, with some farmers even using things like old farm gates to contain the branches. The natural successor to this was the wooden framed harrow that first had wood then iron tines. The drawback to these was that they were light in weight and had to have stones fitted on them for maximum penetrative effect.

By the 18th century the common form of harrow consisted of anything with up to three separate sections of wood framed harrows fitted with iron tines, placed side by side on a common draw bar, but because they were made in a square shape harrows had a major drawback in that the tines all followed the line made by the front ones, hence reducing their effectiveness. Early in the 19th century, due to this defect the shape of the harrow had changed into a rhomboid form.

Even so precautions had to be taken to ensure that the harrow did not wander during use. To avoid this it became the practice for two to be fitted together with coupling hinges and with each attached to a draw bar by

An early root drill with ridge rollers fitted as part of the drill.

a hook and chain. With this system no two tines covered the same piece of ground and many saw it as the ultimate in harrows.

In the beginning, harrows of this type were made with wood frames but gradually versions appeared all in iron. The use of the rigid frame left little room for improvement and even today there are harrows that basically take the 19th century form. In 1839 the first patent was given for a zigzag harrow, with tines at each corner of a parallelogram, and this type is still in use today. Again the complete harrow was made up in sections flexibly linked to the draw bar, and was particularly effective in covering the maximum amount of land.

With the arrival of steel another improvement for the harrow came in the form of spring tines. These were of a curved shape and the theory behind their design was that as the tine was pulled over the ground, the natural forces would try and pull it straight, hence causing it to spring into the clods as well as ride obstructions.

Another form of harrow was the chain link implement. This was developed by James Smith, the man who did so much to bring about a practical mole plough. Designed for the purpose of covering over grass seed, it was made up from a large number of iron discs that tapered to a sharp serrated edge and arranged within a chain web. Simple in the extreme, the chain link harrow has proved effective in use and there are few farmers today who do not have one among their implements, particularly if they need to remove dead grass on grass land.

Agricultural designers however, were not satisfied and continued to search for the ultimate in the harrow. To this end they turned their ideas towards making the tines revolve, and so moved to the disc for cutting through the lumps of soil. The first such development came from America where a patent was taken out in 1847 for a disc type harrow that could be attached to a plough and slice through the soil as it was turned over by the mouldboard. This was the starting point for the disc harrow that we know today.

A simple but ingenious attempt to remove some of the labour from potato planting. With this 'dibber' the holes are made in the ground at required distances apart by means of the spikes on the outside of the wheels, the whole equipment being drawn round the field by a horse.

An old fashioned woo framed cultivator used with a horse. This would be employed before the potatoes were planted to give the required depth of tilth.

Improvement followed rapidly until we had the true disc harrow, a large number of sharp steel concave discs mounted about six inches apart along separate axles set slightly at an angle to each other. To counter any sideways reaction as it is dragged through the soil, each set has the discs set in the opposite direction. Provision was made for the angles of the axles to be varied, giving controlled pulverisation, with the lightest work coming when they were running in line with the pull, and the heaviest when the angle was at its maximum. A seat was fitted for the operator who could make adjustments as the work progressed according to his needs. While the disc harrow evolved during the age of the horse, it proved equally suitable for use with the tractor. With this form of motive power double sets of disc harrows were used, arranged in tandem one behind the other, with the concave discs of the second harrow in reverse order to those of the first. This ensured that the soil was crumbled and thrown to one side by the first discs, to be further crumbled and returned by the second.

A three row root drill which could be pulled by hand or a small pony.

Another form of harrow which survived from the Middle Ages was the 'brake harrow'. Intended for winter use on heavy land that would have defied the normal harrow, it was made as heavy as possible to secure the greatest penetrating action, and in fact was the forerunner of the cultivator, for like the cultivator, the brake harrow was designed for deeper penetration into the ground than the normal harrow.

It was in Scotland, towards the end of the 18th century, that the move from harrow to cultivator was first made, when some farmers began to use an implement with sharpened iron tines designed to penetrate and stir the soil. The ability of the cultivator to 'grub up' weeds earned it the name 'grubber,' a name adopted by John Finlayson when he brought out his self-cleaning cultivator in 1820. This was the first cultivator to be made entirely of iron. It had nine tines arranged in two rows of five and four, designed to the shape that Finlayson saw as ideal for a self-cleaning implement. What he did was to make the tines curve up from the cross member to which they were attached before curving down to ground level. In use weeds and other things that could have clogged the implement were brought up to be swept over the top and cleared away. Three wheels were fitted to the iron frame, and it was the single front wheel that controlled the depth of working. A farm worker walked at the rear of the cultivator, and through a long lever that passed from the front wheel to a type of depth regulator at the back, altered the position of the front wheel, causing the front of the cultivator to rise and fall, hence regulating the depth the tines worked in the soil.

Twenty years later, in 1840, another cultivator, or to give it the correct name, 'Biddell's Scarifier', (a word synonymous with 'Grubber' and 'Cultivator) won for its inventor, Arthur Biddell, the Gold Medal of the Royal Agricultural Society. Two rows of chisel-pointed tines were fitted to a frame carried on two large wheels, and with a smaller front set took the pull from the 'whippletree', or bar that was used in agriculture to keep the draught chains apart. For use on different land, or for other farming requirements, the tines were made removable so that different types could be fitted.

An early type of horse drawn cultivator.

One of the many variations which were to be found on single row seed drills. The drive from the front wheel to the hopper drove the metering arrangement which controlled the fall of seed.

Biddell's scarifier was largely instrumental in bringing the cultivator to popularity, for it was claimed that this machine could break and stir up to eight acres of land a day, and as the second half of the 19th century came, the cultivator was an established part of the farming scene. In some places it replaced the plough in the preparation of the ground, particularly for root crops. It had the benefit of being adaptable as regards the type of land it was used on and could produce a satisfactory seed bed. In the autumn the cultivator was used immediately after harvesting had finished, stirring up the land to allow air and moisture to penetrate before ploughing began.

Later cultivators were fitted with spring tines that were far more effective in pulverising the earth as well as being less prone to breakage. The

A cup feed drill adapted to work from the tractor.

driver, with the horse-drawn implement, walked behind using one or more levers to make his adjustment for depth of work.

With the arrival of the tractor, the cultivator became a stronger and larger implement with a lifting mechanism that could be operated from the driving seat. These later designs were intended as dual purpose implements through interchanging tines. Chisel pointed tines would penetrate deep to break up hard land, whereas the same tines, adjusted to lie just below the surface, were excellent clod breakers. Wide shovel points to give a hoeing action became available, as did broad cutting tines, or shares, for clearing thistle and other surface weeds.

The war against weeds has always been a perpetual battle for the farmer. Left to themselves weeds will get out of hand, shutting out the light that plants need and stopping both air and moisture from reaching them. So weed control has always been an essential part of farming life.

Until the 18th century it was back breaking work done by hand, either cutting off the tops with a sickle before pulling the weeds up by hand, or

A sugar beet field. The crop is being thinned by removing every second plant where the growth is too thick. Work which is still done in the sugar beet field today.

using the more effective hoe. It was an outdated process that made large demands on the farm's labour force. Yet when Jethro Tull offered the farmer a practical form of horse hoe in the 18th century, it was practically ignored.

Tull also gave agriculture its first real seed drill, to sow rows of seeds evenly along the furrows, and devised the horse hoe to clear away unwanted weeds in the space between the plants while at the same time aerating the soil around them.

It is strange to reflect, that while the merits of hoeing were never in doubt, it was Tull's attitude to sowing in a straight line as opposed to the scattering of seeds to every corner of the field that caused controversy and mistrust in his horse hoe.

Farmers brought up on the time honoured 'broadcast' system argued that sowing in rows was wasteful, because a field was only sown to capacity if seed was scattered all over it.

Tull's hoe had been fitted with a 'share', in some ways resembling a plough, but later ones were fitted with a wheel and a varying arrangement of hoeing tines. These early ones, however, still only catered for working between one row of plants.

In fact it was left to a clergyman, the Rev. James Cooke to bring out a horse hoe that could work between a number of rows at the same time, provided of course that the seed had been sown at even distances apart. The implement was in fact a dual purpose one, being a corn drill, and also with the seed equipment removed and hoeing tines fitted, a suitable tool for working between the crop. This was an idea that later manufacturers took up; a multi-purpose drill and hoe which ensured that the seed was planted at the correct distance apart for later hoeing.

Horse hoeing was something that could not be rushed, for with those early implements it was vital for the driver to hold the horse steady, any deviation and the hoe would be cutting through the growing crop rather than the weeds in between. It was the swing steerage hoe that got over the difficulty caused when a horse stepped out of line, and it was this form that was in use until the arrival of the tractor.

Basically the swing steerage hoe was two units. The front unit was a carriage frame with wheels and shafts for the horse, with either chains or rods connecting it to the actual hoe section at the back. The idea was that the chain connections gave a flexibility that enabled the hoe to be kept between the rows even when the horse stepped out of line. The axle of the front section was shaped so it would clear the growing crops while the wheels could be adjusted to fit between different row sizes. Two forms of tines were made for the hoe, the common 'V' shape that worked down the centre of the rows between the plants, or left and right hand side tines that could be adjusted in pairs to hoe on either side of a growing plant.

Today the hoe has adapted to the age of the tractor, but even so, in the case of sugar beet, 'thinning' of the crop often entails hand working in exactly the same manner as the days prior to mechanisation on the farm.

The other implement of cultivation is the roller, which gradually replaced the old hand system of clod breaking until, by the 18th century, it was found on most farms. The earliest rollers were made from materials that were close at hand, like a section of a tree trunk, although wood was never a suitable material because it wore rapidly and was never really heavy enough to work on land without the addition of stones or other weights superimposed upon it. So the simple cylinder of wood with its axle and shafts was fitted, with a box structure on top for the carriage of whatever was to give it the extra weight. Another possible addition was a form of scraper to clean mud from the roller surface.

Early farmers then turned to stone, and while it was suitable for weight, it had a tendency to split and chip. So the breakthrough in roller design had to wait for the time when cast iron became available. Rollers in this material usually had a diameter of about 2', the thickness of the iron deciding the actual weight.

In the beginning these rollers were made in one long length. Then the divided cast iron roller came into being. It was of similar size to previous rollers, about 6' in length, and 2' diameter, but it was split in the middle to give two rollers on one axle. The benefit came when the roller was turn-

A hand root drill, used to fill in any spaces where the seed had not germinated.

ing, a manoeuvre made much easier because one roller could turn forward and the other backward on a tight turn.

The next innovation was the Cambridge or 'ring roller', which took its name from the inventor, W.C. Cambridge, who in 1844 developed a roller that had iron discs threaded on to the same axle. There was a projecting ridge round each disc, and the axle holes were made off centre to throw the discs out of line as they turned. This gave the type of irregular motion over the ground which was effective in use, and also made the roller self cleaning.

A further type of disc type roller was patented towards the middle of the 19th century by W. Crosskill, and this turned out to be an implement which was very successful in breaking up the clods of earth. The 'Crosskill' roller had a number of serrated discs with sideways projecting teeth that cut into the earth and broke up the clods efficiently and easily. Of necessity a pair of road wheels had to be supplied with this roller otherwise it would have been as effective on the roads as it was on the land! Both the 'Cambridge' and 'Crosskill' rollers have survived to play their part in modern farming, together with the ordinary iron roller.

In the days of broadcast sowing when the seed was cast by hand over all the field, 'ridge' rollers could be taken over the field after ploughing to press seed drills into the earth to enable that form of sowing to be as effective as possible.

A similar roller, and an essential implement when planting such crops as turnips, was the 'turnip roller'. This consisted of two concave cast iron rollers mounted on the same axle, and as they were rolled along the ridges of ploughed land it compacted the soil on the ridge and also preserved the moisture content of the land. The next development here was to combine the rollers with the turnip seed drill. Today similar rollers are used in the cultivation of the turnip but they are now to be found at the back of the turnip drill.

In the early days of farming man had no alternative but to cast his seeds by hand over the land he had prepared to receive them. 'Broadcast' sowing, as it was called, could be done with either one hand or two, and the seeds would be contained in either a basket hung from straps round the neck, or a cotton sheet, folded and draped round the upper part of the body. It was something that called for skill of the highest order. The sower would take a pinch of seed between thumb and finger and 'broadcast' it, his hands in step with his feet. Such was the skill of the farm worker of that age if, when the crop came up, he had missed even a square foot in the whole field he would be the general object of ridicule. Although it was a primitive method, 'broadcast' sowing survived into the latter part of the 19th century, and in certain parts of the country, for crops such as clover, it was practised during the First World War.

The other traditional method was to use the 'hand dibber,' nothing more than a wooden handle with a blacksmith-made iron 'dibber' about 2″ in diameter which came to a point and could easily be pressed into the ground to make the hole for the seed. In use a man with a dibber would walk backwards along the furrow slice making holes for the seed which was dropped in by the person following, often a woman or child. This method had the advantage that the seeds were planted in straight rows making it easier for the hand hoe to kill weeds and promote growth generally. Seeds that had been 'broadcast' sown blanketed the field and made hoeing impossible, often leading to a smothered and under-nourished crop. Even when a mechanical aid was brought to sowing, it was a device intended to ensure total coverage, even by the inexperienced sower using the 'broadcast' method. This was the 'fiddle sower,' which came into being during the 18th century to replace, to a limited degree, hand sowing. It took its name from the fact that it was operated in a similar manner to that of drawing a bow over a fiddle, and consisted of a canvas bag leading to a wooden box or other form of seed container, through which the seed was fed to a disc that rotated to the right and left, scattering the seeds in a wide arc as the user walked slowly over the field, moving his bow first one way and then the other. Another mechanical broadcast sower used a small crank handle to actuate a gear chain and so turn the disc that released the seed.

It was not only the 'broadcast' system that received the attention of man's ingenious mind. The 'dibbing' principle also saw the development of other various improvements. Some succeeded but others, like a board drilled with holes at regular intervals for the dibber to be pushed through, being stopped at the correct depth by a shoulder, proved too tedious in use and fell by the wayside. The double dibber was the obvious progression as farmers sought to speed up the work of their sowers, and the idea of using a rotating wheel fitted with iron dibbing points around its circumference was also tried.

The drawback, although this form of implement successfully made the holes, was that, with a single wheel the operator working behind it was treading in a direct line with the holes he was making, and there was the danger that he would accidently destroy them. In an attempt to overcome

A 19th-century seed fiddle which operated in a 'fiddle' action through the handle at the front.

this, experiments were made with a double wheel dibber, but the seed drill was ready to make its appearance, so the dibber in all forms fell into disuse.

When it appeared early in the 18th century, the drill did not receive universal acclaim. For, as Jethro Tull was to discover, the idea of planting such crops as corn in rows, was something that fell on deaf ears, and even after his death in 1741, the old time honoured methods continued to be used. Tull's seed drill cut channels for three rows of seeds which were fed from a hopper into a revolving cylinder, from which they were led into tubes before being dropped into the prepared seed grooves. The drill was light enough to be pulled by one horse and had two large wheels at the front and two smaller ones at the rear. The front wheel axle had a seed hopper attached to it and operated the centre seed dropping system, while the rear wheel axle carried two separate hoppers each with their own dropping system. Three coulters were also fitted to the drill to cut the grooves for the seed.

Inevitably other drills came after Tull's death, and led to two distinct methods of regulating the amount of seed that was released from the drill. These were the 'cup feed system' and the 'force feed system.' In the cup feed drill, seed from the hopper was fed into tiny cups fitted in rotating discs, which at the top of their turn let the seed fall from the cup into a funnel that directed it through the coulter tube to the ground. In the force feed system the seed is controlled by toothed wheels, with it running between the teeth before passing into the funnel and coulter tube. These same systems are in use today, but with the force feed being divided into external and internal.

While forward looking men carried on the work started by Tull, acceptance of the drill in place of the traditional planting methods was slow. Farm workers saw the advent of mechanisation in this aspect of agriculture as a threat to their livelihood, and farmers faced with opposition from their work force chose to ignore the drill. It was not until the 1849s that the seed drill began to find its true place in agriculture. Among the first of the many different types was the 'broadcast sowing drill'. This was simple in the extreme but effective for the type of sowing. It had a wide, narrow seed box, mounted across a type of wheel barrow frame. The sowing was done by fitting a spindle to run the length of the seed box, which had either stiff brushes or iron teeth fitted at intervals along it, and made to revolve through a chain and gear action from the single front wheel. As the brushes revolved the seed inside was agitated and thrown out through perforations along the back of the box. These could be adjusted to control the amount of seed released. So efficient was this form of implement that it survived until recently, particularly for the sowing of grass or clover seed.

One aspect of the 'broadcast sower' which had to be considered by its makers was the length. Drills up to 16′ came into existence and created a problem when it came to moving from farm to field. In some an arrangement was made for the seed box to be swung along the length of the frame, while in others it was hinged into three sections for transportation. In the case of the single row seed drill many forms came into being. In fact the simplicity of the implement which often represented a wheel barrow form, meant that it could even be made on the farm itself, so inevitably the style varied from area to area and often from farm to farm.

Swing steerage, similar to that used in the hoe, was brought into use with the drill, giving the operator the same control over his implement irrespective of the actual path of the horse. Facilities were made for adjusting the coulters and seed dropping mechanism to control the distance between rows, and as men progressed with the drill they gave the farmer an implement that could combine seeds and manure, ensuring that a fertiliser went into the ground at the same time as the seed.

Today it is the tractor drill that is used on the land, and while they are more advanced and cover a larger area at one pull, the basic cup or force feed mechanism is still used for delivering the seed.

Fertilising, either by means of natural manure from the farm yard, or artificial manure made in the factory, is necessary to enrich the soil and help crops grow, and is essential to efficient farming. Modern agriculture has many mechanised aids for this type of work, but until well into the 19th century it was done by hand and the manure cart. Natural manure could be used either in its solid form or as a liquid. The former was carried by cart and distributed over the field by hand, often by the women of the village, helping out in the fields. Liquid manure, made by directing rain and other waste water into a holding tank, was carried by the manure cart, consisting of a cask holding anything up to 150 gallons, supported between shafts and fitted with an axle and two large wheels. The cart was filled via a funnel through a hole in the top, and the liquid was let out through a perforated tube running along the back of the cart. It was a wasteful system, the manure being spread all over the land as opposed to being directed to where it was needed, on the actual crops. Liquid manure, like every other aspect of farming, came under the scrutiny of the designers concerned with adding to the efficiency of the farming scene, and in 1885 Thomas Chandler of Aldburn brought into being a manure cart that enabled the manure to be deposited onto the growing crops without waste. It operated through a series of small cups on an endless chain in a principle similar to a dredger working on the ocean bottom. Power from the implement's wheels drove a series of gear wheels which activated the chains, and the cups filled with the liquid before discharging it into delivery pipes through which it flowed to the ground.

With the arrival of the man-made fertilisers, liquid manure fell into disuse, but in recent years, with the idea of organic farming gaining in popularity, many farmers have reverted to the use of manure found on the farm yard and, converted to liquid manure this has been used during the past ten years or so, particularly on the dairy and pig farms. Here liquid manure, or 'slurry', is stored in holding tanks and when needed, sprayed on to the land. There are, understandably, environmental problems caused by smell where this system is adopted.

4·The Barn and its Implements

For the arable farmer, the most important building on the farm was the barn. This was where the sheaves from the harvest field were stored and the scene of threshing operations; the separation of the grain from the chaff during winter. The barn is another example of how agricultural buildings as well as implements evolved to fulfil a distinct farming need.

The barn was in every respect functional. There were storage facilities at each end and in the centre were two sets of double doors so that the harvest wagon could come in through one set and leave, when empty, through the other.

The barn was usually sited in the central area where the threshing was done, and the floor area was specially prepared. The ground would be dug out to a depth of about nine inches, before being filled in with a layer of dry sand and gravel, ready for a topping of clay mixed with cow dung and chalk or something similar. Each layer was rammed down and the whole allowed to dry out whereupon any cracks that appeared were filled in, to leave a threshing area that was smooth and as hard as concrete.

In some of the larger farmsteads there would be big barns fitted with two threshing areas, two sets of double doors, and of course double the amount of storage space. Often there would be two barns, one intended for the wheat that would be sold, and the other for the oats and barley that would be used to feed the livestock.

A barn of some sort was essential ever since man started to wrest a living from the land, but evolution and change made even this aspect of the farming scene redundant. For as mechanisation progressed in agriculture the old fashioned flail gave way to threshing machines which were worked first by steam then internal combustion engines, not in the barn but outside in the farmyard. By then the corn was being stored in ricks, so the barn, for the arable farmer, had outlived its usefulness.

In one respect however, when the combine harvester arrived on the scene, mechanisation turned things a full circle. The corn rick had served a dual purpose, protecting the crop from the weather but at the same time allowing the wind and atmosphere to dry it. Now with a machine that delivered the grain threshed and cleaned in the harvest field, the farmer, needing to dry it to prevent the formation of mould had no alternative but to turn to the engineer for a mechanical drying system, which, in the main

One of the many forms of chaff cutter to be found in the barn. Straw was pushed along the trough into the housing with the cutting knife.

was built into the old barn, so that once again it became a prominent building for the arable farmer.

Before the arrival of the threshing drum the most important function of the barn was to provide a place for threshing. With wheat, the universal tool was the flail, but for a barley crop what was needed was a tool to remove the long spines or awns, and for this purpose the 'hummeller' was used. Early farm workers accomplished this by working a wooden roller backwards and forwards across the barley, a task made far easier when the hummeller made its appearance during the 18th century. Hand hummellers, made locally by the village blacksmith, varied in shape and size but each followed a basic pattern. A shaft some 3′ in length and similar to a spade shaft was joined at the bottom to the hummeller head. This was a square iron frame inside which was fixed a series of iron bars that formed separate partitions between 1″ and 2″ square.

63

A chaff cutter working from the threshing drum.

In use, the barley was laid out on a rough, uneven floor and the hummeller lifted and brought down on the crop, which was then turned with a shovel in the air stream to help clean it. A constant pounding action with the hummeller was maintained which broke off the spines or awns and made the grain suitable for sale.

The next adaptation was to use the principles of the hand hummeller for a roller hummeller that could be pushed backwards and forwards across the barley. This was done by arranging metal discs along an axle and joining them with thin iron bars to divide the tool into small compartments as on the hand hummeller. By the 1830's a hand operated hummelling machine found its way onto the market. Standing 4′ high, it had a central vertical shaft that took the barley through a series of pins passing between each other, and hence removed the awns and delivered 'hummelled' barley through a spout on the side.

There were other barn tools that were as vital to the operation of the farm as flail and hummeller. Riddles were necessary in the work of cleaning the grain, and removing heavy foreign matter from corn of any kind. In the main these were made entirely of wood, but a few were of wood and wire. Many farmers believed that in the hands of a skilled man the latter was the best. Riddles were usually about 2′ in diameter and 3″ deep, with beech being the most widely used wood for the rims, although fir or oak was also used. For the withes at the bottom, if metal was not used, then either fir or willow would be chosen, although there were those who saw American elm as perhaps the best. The mesh size depended on the use to which the riddle was put. A wheat riddle would differ in size from a barley riddle or an oats riddle.

When the grain was being measured, accuracy was vital. To this end the Weights and Measures Acts saw the imperial bushel, a measure of capacity containing eight gallons, as the standard for either wheat, oats or barley. Before it could be taken into legal use it had to be stamped by the authorities to show that it contained 2218.19 cubic inches volume, although it could vary in shape. Normally this weight was the work of the Cooper, made of oak and hooped with iron in the form of a barrel with a narrow neck and broad base. An essential was that the neck was of a shape that could be wholly or partially inserted into a sack so that none of

A threshing drum and chaff cutter overtaken by progress!

the contents was lost. Also demanded by the Weights and Measures legislation was the corn strike, for sweeping excess grain off the measure. Officially this had to be in the form of a wooden roller 2″ in diameter, but many farmers, in the belief that a roller only succeeded in compressing the grain, used a sharp edged strike which they worked in zig-zag fashion across the measure.

Other tools for this part of the farming operation were the wooden corn scoop or corn shovel, as well as various weighing machines, sack lifters and barrows. The sack lifter was simple in concept yet valuable in use. The small wooden handle fitted comfortably into the hand and with twin hooks on the end took much of the strain out of lifting sacks. Again the sack barrow was a tool that could be made on the farm, consisting of two side pieces curved at each end so they could rest easily on men's shoulders and with three wooden cross pieces on which the sack rested. By the use of this hand barrow with a man at either end, weights could be moved that

A plate mill of the type to be found in the barn. Two serrated metal plates would be used, one stationary and the other turning by hand or by a stationary engine.

Various types of corn measures together with riddles for separating the corn. In the foreground is the hummelling roller used to separate the barley crop.

would otherwise have proved to be difficult in the extreme. In fact some of the barn tools of yesteryear were so simple that it would be easy to dismiss them as unimportant. A typical example is the broom that was used to sweep up any corn that had been spilled. The 'besom', to give it its correct name, was made by those who used them from birch cuttings bound to a wooden handle. It was primitive, but it was functional and without complications did the job intended for it, a factor that ran through almost every piece of farm equipment.

Perhaps one of the most ingenious tools was the grain sampler that came into use about 1870. A conical shaped container was fitted with a screw-threaded shaft. In use the 6′ shaft was unscrewed so that the container mouth was open. Then it was pressed deep into the grain so that some would fall into the container which would then have the shaft screwed down, sealing the mouth and enabling a sample to be taken from the bottom of the heap.

As well as being the place where men toiled over the threshing, the barn also housed the machinery necessary to prepare foodstuffs for the farm livestock. Many were hand tools, like those used on the turnips fed to sheep and other animals. The first stage was to cut off the leaves, and to this end a knife in the shape of a bill hook with a spike on the end came into its own. The spike could be thrust down into the turnip avoiding much back-bending on the part of the user. Simple but effective was the tool used for cutting the turnip into small pieces, for this was nothing more than a spade-type handle to which was fitted crossed metal cutting blades.

Straw which had to be cut into short lengths to feed the animals was something else that originally was sliced by a man using a knife, but the actual 'chaff cutters' as they became known, were quick to develop. In the beginning they were nothing more than a 'cutting box', a wooden trough open at both ends and supported on four legs. A knife-edged lever was fitted at one end, enabling the farmer to push his straw or hay through with one hand while he cut it into pieces with the other.

The first addition to this was a foot-operated control that held the straw down while the cut was made, but gradually designers came up with a chaff cutter which had its cutting blades fitted into a 'wheel' arrangement, with the straw being fed through automatically by toothed rollers as the handle fitted to the cutting wheel was turned.

Such was their dependability that even today chaff cutters of this type are in use, although powered now by either electricity or the tractor. In the age before the internal combustion engine appeared on the farm, manufacturers were turning to steam to give farmers with large numbers of livestock to feed, a machine capable of keeping pace with their demands.

Root crops, such as turnips, were also receiving the benefit of manufacturers' ideas. Although the first device for cutting the roots to a size that made them suitable for sheep would have been similar to the straw cutting box, the two showed little resemblance when further progress followed, for designers turned to a lever system whereby a shaped wooden block was brought down on the root crop which had been positioned in a container with sharp knife blades in the bottom, shredding the root into slices.

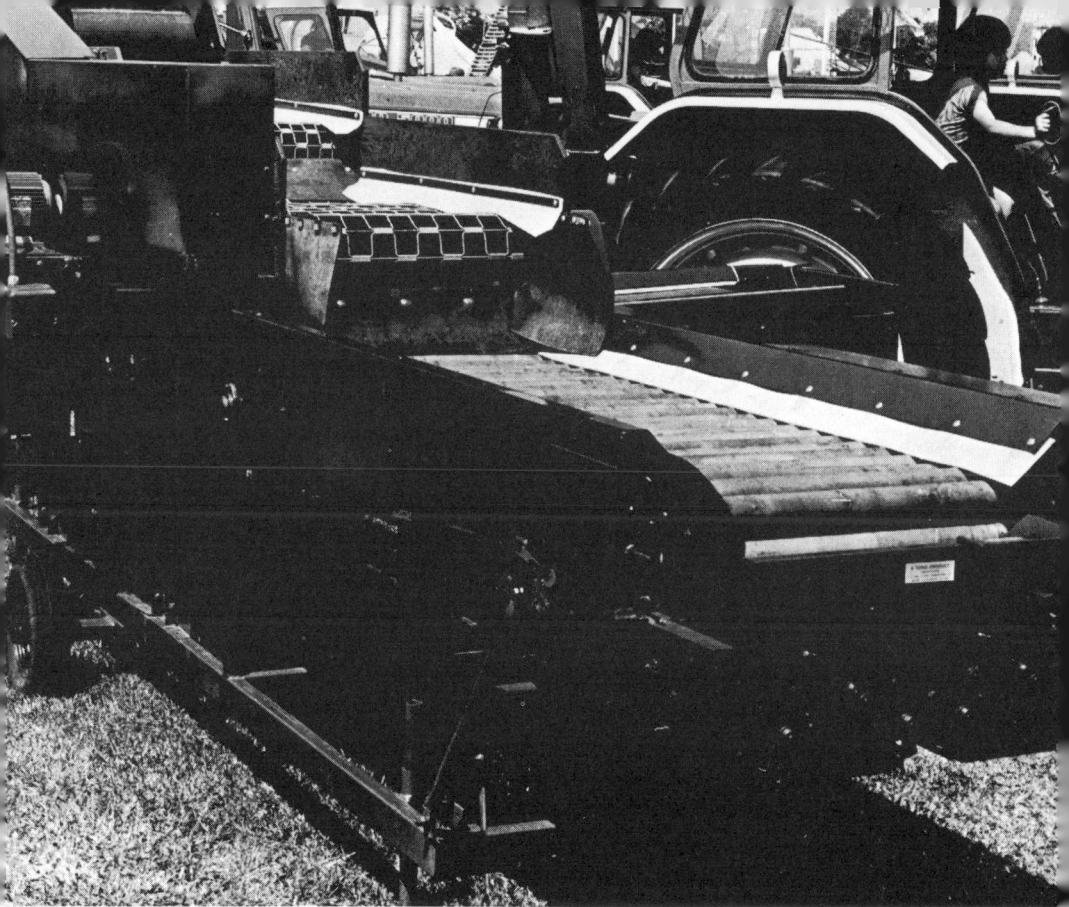

A modern potato riddle for grading potatoes into their respective sizes.

The next development was perfected by a Banbury implement maker, Mr Gardener, in 1839. He utilised a series of staggered cutting knives on a roller, the roots being fed in through a hopper and effectively sliced without any waste when the handle of the cutter was turned. An earlier variation of the rotary cutter used a design similar to the chaff cutter with two thicknesses of cut available depending on which way the handle was turned.

Another aspect of food preparation was milling. For although some animals could eat whole grain, it was the practice to grind it to aid their digestion. Three types of 'mill' were available for this task. The hammer mill, which as its name implies used hammers fitted to a rotor which turned at high speeds. The crushing mill ground finer than the hammer mill, using a 'windmill type' stone action through iron plates. Finally there was the roller mill with mangle type rollers into which the grain was fed, their distance apart being adjustable, hence giving variation to the extent of 'milling' applied.

Another form of grinding machine was the 'cake breaker'. Oilcake, the end product from linseed and rape seeds after the oils have been extracted, was first recognised as a manure. However, in the later part of the 18th century it came into use as a feedstuff capable not only of fattening sheep and cattle but producing an enriched manure from the beasts which had been fed on it. For convenience of transport and storage it was produced in large thin slabs, but when these reached the farm they had to be broken by special machinery. For this purpose 'cake crushers' made their appearance in various forms during the 19th century. Rollers were the central requirement, with the first 'breakers' having only one pair, fitted with spikes. In use, the cake was dropped into a hopper that fed it to the hand-turned rollers. Considerable pressure was needed to crush the cake and this was obtained by gearing down the hand-turned wheel, so bringing greater power to the roller action. These early machines were in some respects dual purpose ones, for the roller action resulted in both lumps of cake for cattle feed and a fine dust that could be used by the farmer as manure.

Later developments, particularly after the arrival of steam, had additional rollers fitted, and according to the thoughts of the manufacturer when he was considering efficiency, these would be either spiked, notched or grooved. Cake breakers were used on the farm into the 20th century, but the advent of concentrated feed stuffs in the form of cubes or nuts has rendered another piece of farming equipment redundant, although many survive to remind us of the lost age of farming.

Bones were another material used in agriculture, and these needed to be crushed before they could serve their purpose as a fertiliser. Bones were of great value during the 18th century and later, being used for such diverse purposes as cutlery handles and buttons. Because of this, if the bone meal fertiliser came from a factory it was the smaller, second class bones that were used. Bones however were readily available to the farmer from the carcases of his own cattle, so a bone breaking machine on the farm was a valuable addition to the range of barn implements.

In general these machines used a series of rollers arranged in pairs and with their surfaces either grooved or fitted with teeth. The machines were powered by animal or steam and the roller gaps were arranged so that minimum pressure was applied at the beginning to avoid choking the

A typical oil cake crusher. The oil cake was passed through rollers to crumble it into a state fit for feeding to the cattle.

machine, but as the bones progressed through it, the pressure was intensified. The bone crushing machine appeared on the agricultural scene towards the middle of the 18th century when the fertiliser value of crushed bones was beginning to be appreciated, and survived for about a hundred years until commercially prepared fertilisers began to make their appearance.

Another source of animal feedstuff in those areas of England and Scotland where it was readily available, was gorse. The limiting factor with this was of course its spikes, and because of them the gorse had to be crushed to a pulp before being fed to cattle.

Originally, until the introduction of the crushing mills by about 1820, this was done with the 'win bruiser', a tool very much like the barley hummeller. The shaft of this was made of a heavy bulk timber so that the maximum weight could be applied to the gorse crushing.

In Scotland, where the gorse was available in bulk, a permanent gorse cutter powered by a horse and utilising an old mill stone was in use since the early 19th century. An axle was fitted through the centre of the stone. One end was loosely attached to a central pole and, pulled by a horse, the mill stone ran on a circular flagstone path on which the gorse was laid.

Root slicer for cutting up such crops as turnips for feeding to sheep.

When it arrived in England the gorse mill was a portable machine intended for use in the barn or yard. It was effective in operation and consisted of rows of iron teeth so arranged that when the external handle was turned one series rotated close to the others so that as it passed between them from the loading hopper the gorse was crushed into the desired pulp to make it an admirable animal feedstuff.

Another use for the gorse machine was to pulp turnips and other roots needed to feed the farm stock, but this was accomplished far more satisfactorily by the machines designed for that specific task, the root pulpers and slicers. Root pulpers worked on a similar principle to the root cutters, but the knives could be adjusted to either pulp or shred. Pulped food was used when animals were young or ill, gradually progressing to sliced foods, and again in the final stage in fattening up cattle either for market or slaughter.

In places such as Devon an important piece of apparatus on the farm was the apple mill used in the making of cider. These, like so many other implements for rural use, were primitive in the extreme, consisting of nothing more than a wooden roller to which nails had been fitted to grate the fruit as it dropped down from the hopper. The grated apples were then beaten in a stone trough to pulp them ready for the press.

Inevitably, manufacturers turned towards this form of press, and an improved version with two sets of rollers, one spiked to draw the fruit into it and the other smooth to reduce the fruit to the desired pulp in one action, came onto the market. Today, as a place for cider making the farm has largely given up, and the process has become one for the factory.

The barn was also the place where the potato crop was cleaned and graded. In the beginning the cleaning was done by hand, washing the crop in a barrel of water. The larger potatoes would receive individual treatment, but the smaller ones would be put in a basket and dipped into the water.

By the early 19th century this aspect of the operation had a mechanical aid in the form of the potato washing machine. In the beginning it was nothing more than a wooden cage that revolved in a trough of water with the action of the potatoes rubbing against each other being sufficient to clean them. When this had been done they were dropped out into a slotted rack to dry. As man's knowledge progressed so the potato washer retained its basic principle but was fitted with heated fans to dry the crop, so leaving them ready for bagging.

Potatoes are one of the farming crops that have to be graded; the sorting of the potatoes into large, medium and small sizes. The smallest were suitable only for feeding to the farm cattle, the medium ones were used for seed purposes and the larger ones sold off. In the early days grading was accomplished by using a double riddle. The larger mesh at the top allowed all but the large potatoes to fall through, with the second mesh sorting out the seed potatoes from those that would go for animal feed. A later system used a revolving cylinder which was set at an angle, with varying spaced bars. First the small ones fell through, followed by the seed potatoes further along, with the largest working through to the end of the cylinder for bagging.

Possibly the most unusual of all barn machinery was the 'smut machine.' Smut was a disease common in corn and took its name from the soot-like powder that appeared on the ears of grain. Obviously it would be difficult to sell the crop in that condition so men turned their ingenuity to removing it. The method they chose was a fine gauge cylinder arranged so the diseased corn was fed through it and brought into contact with rows of rapidly revolving brushes that removed all traces of the 'smut' and left the corn clean and shiny.

Disease has always been a problem for the farmer. Today the seed comes ready treated, but in the days before this breakthrough by scien-

Inward sloping wheels were employed to squeeze the beet from the ground.

tists and chemists, the farmer was still prone to outbreaks among his crops and had to prepare his seed against it. This was another task that was done in the barn, usually just before the seed was sown. The seed was put into a wicker type basket and immersed in a solution of copper sulphate. Then it was held over a trough to drain away the surplus moisture, leaving seed that was coated with copper sulphate and so less susceptible to many diseases, including 'smut'.

Today the machinery used in the barn has changed with the advent of the internal combustion engine. Potato graders, although giving the same end result of large, medium and small samples, are driven by their own motor, although such implements as the chaff cutter, root slicer and pulper are still playing their part in animal feeding. In fact the modern barn, if we still give that name to the prime building housing the end product of the harvest, is a tribute to the mechanised peak that agriculture has attained, for grain, in its harvested condition, even though some natural drying may have taken place, will have a moisture content that is far too high for storage. So, if the farmer is not going to risk losing his grain through mould or fermentation from the generation of its own natural heat, it must be put through a system that will extract most of the moisture either before or during storage. The method used depends upon whether the grain is to be stored in bags or in bulk. Grain that has been bagged uses a platform drier which consists of a concrete floor with gratings fitted, and with air ducts below them. A fan and heater unit blows warm air through the ducts, hence circulating it by way of the gratings round the bags of grain.

One way of drying loose grain is by means of a ventilated silo drier. Here the large tank, or silo, has a ventilated floor through which warm air can enter, circulating through the grain, drying that at a lower level first but moving upwards until all the grain in the silo has been dried. An effective system is the moving of the grain itself through a series of drying and cooling tanks, before it is taken out of the machine for storing elsewhere or using. This method however is expensive and is used only on the large grain producing farms.

Since the energy crisis, another approach to the drying problem has been to blow cold, dry air through the grain crop. In its damp stage corn generates heat, so by using a fan blower to send this cold dry air through it, the self-generated heat is removed and the corn is consequently dried. It has an energy saving benefit in that only power for the fans is needed, no heating being necessary.

5 · Root and Other Crops

The First World War taught Britain the important lesson that in any future conflict her food supplies could no longer be guaranteed. In fact it was the serious results of the 1917 German blockade of our shores that brought about a re-think in relation to our agricultural policies when hostilities were over. One of the matters that needed urgent consideration was our position with regard to sugar supplies. A factory to process cane sugar had been built in England as early as the 16th century, but cane sugar had to be imported and what was needed was a source of home grown sugar.

Even before the outbreak of the 1914-18 war, a form of sugar was being grown in Europe. This was sugar beet, a root crop similar in appearance to a parsnip, for which much pioneer work had been done in both France and Germany, developing through breeding and selection, a good sugar producing root crop from a form of mangel. After those war time experiences, the value of home produced sugar was not lost on the Government. The foundations of a beet industry were laid and by 1922 it had become an important farm crop, producing not only sugar for human consumption, but animal feedstuffs from the beet tops and the residual pulp received from the processing factory, for beet is the one farming crop that is offered to the farmer in the form of a 'package deal.'

The ruling body is the British Sugar Corporation which operates the 'beet factories' that are a feature of the beet growing areas of the country. When the crop is received from the farmer it is first tested for its sugar content then the sugar juices are extracted from the root. After this impurities are removed and the juices are put through boiling processes to give the refined sugar either in cube form or loose.

Although beet was a late arrival on the farming scene, it still needed a large amount of hand labour for planting in the early days. In fact, although mechanisation has now come to the aid of the farmer, much hand work is still done on many farms which grow sugar beet. For while many drill the seed to a standard of between 6″ and 8″ and hence eliminate the need to work through the field removing plants that are too near their neighbour, some farmers plant at least part of their crop close enough together so that if there is a failure to germinate, they will be close enough so that the missing seeds do not affect the yield.

Potato implement which would have been drawn by a horse. Attached is the ridging plough share used for planting and ridging the crop. In the foreground is the lifter which was attached in place of the ridger to work under the row and lift the crop, letting the soil drop through the bars.

This system, with seed set every four inches, means that when the plants are appearing through the soil hand labour is necessary to work them with a hoe, removing all excess growth and leaving the field with the beet spaced at distances to give the best conditions for growth.

Harvesting also differs from many other crops in that the beet has to be 'topped' before it is lifted. First a horse-drawn lifter went over the field, and men following removed the dirt by knocking two beets together. Then a man with a knife would come along, top the crop and throw it into heaps for the horse and cart that carried it from the fields.

There are two different ways in which the beet itself can be lifted. A pair of inward sloping wheels, 'Oppel wheels', can be used to squeeze the beet out of the ground, or a pair of shares to cause the beet to ride up them may be fitted.

In the past, when the horse began to work in the beet fields it was a two implement operation, but gradually mechanisation brought the farmer the one step harvester which carries the beet along a conveyor with a shaking action to clean it, and then into a storage tank until it is transferred to the trailer to be taken to the farm.

Another root crop that required special treatment, both in sowing and harvesting was the potato, which is far older than sugar beet, having been introduced into this country in the 16th century. However it was not at first seen as a commercial foodstuff, but rather as one of nature's curiosities. Now of course, it is one of the most valuable of all farm crops as a source of food.

Up to the middle of the 19th century potatoes were planted by hand with both men and women working in the fields, dropping the crop into previously prepared ridges. Stephens relates in his 'Book of the Farm', how the potatoes were put into a cart which was taken to the head of the field where they were unloaded with a frying pan shaped shovel. They could either be put directly into the round wicker baskets of the planters or, in a large field, into sacks at convenient points from the planters. Natural dung from the farm yard was also taken to where the potatoes were being planted and dropped into dung hills, when the women, using small three pronged forks, would work it into the drills before the planters deposited their seed.

One mechanical aid that man discovered for use in the potato field consisted of nothing more than a number of iron wheels mounted on a common axle, at a distance equal to the rows and with protruding points which made a hole in the ground to receive the potato.

When the mechanical potato planter arrived on the farming scene it evolved into two methods for dropping the potato. In one an endless chain fitted with cups gathered the potatoes and dropped them into the tubes leading to the ground. Another version had revolving discs to which small forks were attached which impaled the potato before dropping it into the planting tube.

The planter itself can be found in three different types. The manual planter is fitted to the tractor tool bar. Potatoes, like sugar beet, have to be planted in rows spaced to leave enough room for further cultivating and spraying, so the tool bar, provided with spacing clamps, can be arranged so that the crop is planted at the correct distance between rows. In use the potatoes are put into hoppers and dropped by hand down the delivery shutes each time a bell rings.

A semi-automatic planter also needed men to feed potatoes into com-

78

Lifting the beet. This would be drawn by a horse.

partments in a vertically revolving wheel which delivered them into the delivery chute from each compartment in turn. This was both quicker and easier than the manual planters.

Automatic planters are now on the farming scene with a feed mechanism which picks the seed out of the hopper and takes it by a conveyor to the chutes. Additional attachments on all planters are the coulter that makes the groove in the ground for the potato, as well as a means of ridging to cover the seed once it has been planted. In relation to harvesting the crop, up to the 19th century it was a task done by hand. An army of men would work the fields using the potato graip, a fork with three prongs flatter than normal and wider towards the end, before tapering to a point. It was an effective if back breaking tool, turning the potatoes out of the ground where they were collected into baskets by women and children.

The first mechanical aids were the brander and graip for use on the plough. Both used a system of iron bars which were arranged in a 'fan' shape with the narrow end being at the front behind the plough share. In the case of the 'brander' the mouldboard was removed completely, but the graip was used behind the double mouldboard on that type of plough. In use the potato lifters had the soil and crop passed over them, the soil fell through the rods but the potatoes remained on them to be dropped off on top of the ground ready for picking up.

A potato digging implement made its appearance in 1855 and its rotary action laid the pattern for similar implements of the future. The first made by a Mr Hanson, had two small wheels at the front and two larger ones in the middle to provide the power for the gear system that operated the rear mounted spinner. This consisted of a rotating disc with fork tines fitted and arranged so the ends of the tines were able to penetrate about 9" below the soil enabling the potatoes to be spun out of the ground, whereupon they hit a hanging net and were left in rows for the gatherers.

There was one drawback to this type of machine. With a fixed circle of travel the spinners were unable to gather any potatoes laying at the sides of the rows and these tended to be lost to the farmer. A variation to overcome this drawback was designed. In this machine the tines turned in an eliptical path, digging over a greater width and recovering potatoes that otherwise would have been left in the ground.

It was this type that was developed further, producing efficient potato harvesters for both horse and tractor. Other approaches were also tried, including an elevator digger with the potatoes being lifted by mechanical shovels and the soil falling back through a system of sieves, while the potatoes were thrown out at the end of the machine.

The modern concept of potato harvesters however, has moved away from the spinner. The 'Hoover' uses a share on the front to lift the crop which is pushed onto a wide chain at the rear. Vibration of the chain causes the soil to drop through back into the ground while the potatoes are deposited on the ground waiting to be picked up by the farm workers. 'Hoovers' such as this can be worked in series, one behind the other, throwing the potatoes in different directions until they can be collected by the elevator lifter and deposited into the tractor and trailer moving slowly alongside.

Again there are machines that combine the two actions, lifting the crop by mechanical means and feeding it straight into the tractor and trailer working side by side with the harvester. In the field of harvesting, modern science has been able to give the farmer the ultimate in electronic eye control over his potato crop, but at a high price it is a machine for the large scale farmer. For the small farmer the potato harvest still finds the women flocking to the field to pick up the crop in a manner that has been an established part of agriculture since potatoes were first grown as a food in England.

There has always been more to potatoes than just growing them, and in the early 1930's farmers with this crop were in a state of utter confusion. Often they had no idea of the price they would receive until the cheque arrived from the merchant. If prices were high, imported potatoes flooded into the country with the obvious effect on the home grower. It was a position that had to be controlled and to this end the 'Potato Marketing Board' came into existence in 1934.

The board brought stability to the potato grower by ensuring that they controlled the acreage planted in the country, and also by seeing that potato merchants were licensed. In the old days, before the advent of the Potato Marketing Board, the farmer had no guarantee of the financial background of the merchant he dealt with. There were cases of merchants collapsing with a heavy financial loss to the grower, but with the formation of the Board this is a factor that has been removed.

While roots such as beet and potatoes are important crops on the farm, grass is equally essential to the farmer concerned with raising cattle, for it is on his hay yield that his winter feed depends.

The hay harvest is the first in the farming year, and one that is very dependent on the weather, for the hay has to be mown at the time when it is in flower or coming into flower, and the period between this and when it begins to lose its quality is very short, so hay making is often an urgent and worrying time.

In the early days the work was done by hand, men with scythes or sickles doing the cutting. However, the cutting is only the beginning of this farming process, for it is essential for the hay to dry before being stored for future use. The swathe of hay left on the field after the cutters had done their work would only dry on the top, so it was vital that the crop was constantly turned over so that the whole of it received the drying effects of the sun and wind.

Before the advent of mechanisation it was a task often done by women working it with forks and rakes. In the evening, to save it from the full effects of the night moisture it used to be gathered into 'haycocks' to be spread out again the following morning. The turning action that was done by the women was known as 'tedding' and involved throwing the hay into the air, hence opening it up to the full effects of the sun and air.

The problems concerned with hay making, except when grass is converted into silage, have not changed since those early days, except for the arrival of machinery to carry out the necessary work.

With the advent of machinery we do tend to forget the importance of the simple tools like the rakes, which were of such vital importance during the period when the crop was drying. However, in the days when they were so essential in the field, they were in such demand that making them was a rural industry on its own.

The hay rake differed from the normal rake. It had to move the hay but at the same time not dislodge any stubble from the ground. In general it had an ash handle about 8' in length. The head was made of harder wood and had anything up to twelve short wooden teeth with rounded ends fitted to them. They were usually screwed in so they would not fall out when in use, and this system made it easy to replace any that were broken. Although the design varied from area to area, all followed the same pattern in relation to length of handle, angle of head and method of use.

In many ways early attempts to bring mechanisation to haymaking followed the trend for harvesting cereals, using swinging or rotating knives. However, when the first successful mowing machine was produced in 1831 by the American, William Manning, he used the 'reciprocating cutter,' a system that is still in use today.

In these horse drawn hay cutting implements, there was a frame work which carried the two ground wheels that provided the power for the gears operating the cutting mechanism. The framework held the shafts for the horses and a seat for the driver. The cutting part extended sideways and divided the grass into bunches before it was cut with the reciprocating knife. This type of cutter has altered little right up to our modern age although of course it is now driven by the power taken from the tractor. Another form of mower found in the fields today uses a cutting disc action in the rotary mower.

But it was the 'tedder' more than the mowing machine that changed the hay making scene when it appeared in the first part of the 19th century. In the past it had been done with hand labour, women throwing the hay into the air to open it up to the efforts of the sun. It was nothing more than a wooden cage formed with four longitudal strips of wood carrying iron spikes that revolved through a drive from the two ground wheels. As it was drawn across the hay by a horse the wooden cage revolved and the

A common form of sugar beet harvester in use today.

spikes caught the hay and tossed it into the air. Primitive it may have been, but contemporary records show that with the horse moving at a gentle trot it could do the work of fifteen women in a more efficient manner. One drawback was that on uneven land the spikes tended to catch the ground, and of course if used in windy weather there was a danger of the hay being blown all over the field. The division of the cage into two parts and the fitting of springs overcame the disadvantage of working on rough ground, although the throwing of the hay in windy weather stayed with the tedder. Some owners tried to fit canvas or metal hoods, but it was only when the horse hay rake came into use that this objection to the use of the tedder was removed.

The rake that would be used after the tedder to bring the hay into straight swathes had a number of curved tines supported on a wheeled carriage with a seat for the driver. In use it was driven across the field and as the rake filled with hay, a lever was operated that lifted the tines and dropped the collected hay thus forming neat rows.

There were other methods of collecting the hay and one of the simplest used before the arrival of the horse rake was either a rope or a curved piece of wood with ropes at either end which was dragged across the field collecting the dried hay. These were the simplest forms of 'hay sweeps,' others were in the form of a forked scoop, fenced to hold the hay it collected, or what looked like a three sided ladder, locally made, and drawn across the field by horses.

The American Hay Rake that came into Britain in the middle of the 19th century was adopted in many regions of the country. In appearance it was like a large double toothed comb that was pulled over the field, with one set of tines pointing to the front and collecting the hay, while the other set faced backwards. When the rake was full the driver operated wooden handles that caused the tines to move in a rotating action, releasing the load collected and bringing the back tines into action to gather another load.

Another American introduction to the hay field was the 'Keystone Hay Loader,' which could be fitted to the rear of a farm waggon, taking its power from ground wheels while fork teeth rotated to pick up the loose hay which was taken by an endless belt elevator to the waggon.

America was also the centre of development of the baler, the implement that compressed the hay into the bales that we are familiar with today. One baler came into being in the middle of the 19th century but it was too bulky, delivering huge bales that made hard work for those engaged in taking in the hay. It was the larger baler however, that made its first appearance on the farming scene. These were intended for use in the farm yard as opposed to the hay field, and were powered by either a 'horse walk', the means whereby a horse walked round and round transferring power through a large cog wheel to the implement, or by steam from the farm traction engine.

The pick up baler, the implement that now works in the field, picking up the hay and making it into a compact bale, is a modern machine dating from the 1940's.

Another 20th-century implement for the hay field is the 'swathe turner.' The more common form that is now seen in the fields is the 'finger wheel' type. These are large wheels made up of projecting tines coming from a central disc. They are free running and turned by the tines touching the ground, and can be used either for turning the swathe for drying or for side raking, bringing two swathes together.

The wheels which are set obliquely in their frame, overlap one another. The finger wheels are arranged in either fours or sixes, and when used for turning hay for drying are mounted in pairs. If they are to be used as a side delivery rake, getting the hay ready for the bailer, then they will be fitted in line behind the tractor.

In modern farming there is another approach to using the grass grown on the farm, and that is to make it into silage, which means that it is cut and stored while green. This is a quicker method than hay making and more certain because it is cut and can be stored, in its wet state, immediately.

The machine that is used to cut, pulverise the grass and blow it into the accompanying trailer is the forage harvester, an implement which appeared on the farming scene during the later part of the 1950's. Even so, hay making is very much a part of farm life, and while the old haystacks that were once such a dominating feature in the countryside have disappeared since the baler reached its present state of perfection, the old methods of working for the sun and the turning and re-turning of the hay crop, is something which can still be seen.

6 · Steam on the Farm

If there is one part of the farming story that has truly lived through into our own period, it is the age of steam. Through the dedicated work of enthusiasts nearly every aspect of steam as it was harnessed to the needs of agriculture has survived. During the summer months, at meetings large and small it comes to life again to perform as it did when steam was the motive power for every part of farming activity.

Traction engines and the early portables can be seen providing the power for threshing machine or saw bench, while the mighty cultivators devour the soil in the way that was common when steam ploughing was the province of the contractor who would send his engines out on a yearly agricultural roundabout.

Like many innovations however, steam was slow to take the ascendency. Its beginnings, in relation to the farm, could be said to have originated with Richard Trevithick, a Cornishman who could indeed claim the title, 'Father of Steam', for it was he who released designers from the inherent disability of Watt's low pressure engine and boiler, with its huge cylinder and piston, and gave them the modern high pressure engine, which although smaller was capable of developing similar power.

In the beginning Trevithick concentrated his attentions on a steam road vehicle before lack of finance saw him turning to the development of steam railways. It was to the latter's benefit and the former's loss. Yet despite the setbacks and disappointments he experienced with his steam road vehicles, he never completely forgot those early ideas, and by 1812 in advertising a portable steam engine for threshing and grinding corn and other agricultural uses, he was directing steam towards the farm.

The climate of opinion however, was against the introduction of such engines. The country was in a disturbed state. Grain prices were high with a subsequent drop in demand. Farmers as a whole were cash conscious, and although some of the larger farmers bought Trevithick's portable, and found it lived up to all its claims, the smaller farmer was not prepared to make a cash outlay, on an engine that was as yet unproved. So, for a period of some twenty years, the portable engine stagnated. Many who, in other circumstances would have turned their attentions to this new development in the field of agriculture, let the opportunity pass. Progress however, was not to be denied. By the 1830's the wind of change was blowing

86

through the agricultural world. The market had become more stable and those who, only a few years earlier, had held back from manufacture now considered improving the early portable engines that Trevithick had made.

It is often not enough just to have a good product, a man must be able to read the market and grasp the opportunity when it presents itself. One unfortunate who failed in this respect was a man called Howden who, at the 1839 Agricultural Show at Wrangle, Lincs., both astonished and delighted the farming fraternity with an impressive portable engine.

Although small when compared with others of its time, it produced adequate power for farm use, and test reports showed it to be reliable. Many farmers saw it as the answer to their power requirements, and orders flowed in. Howden, however, was not impressed by all this public acclaim. Manufacture, he said, must be kept at a low level otherwise engines would flood the market, the country would be overstocked and the maker who had invested in premises and equipment ruined. A fortune evaded him due to his lack of business acumen. While he turned his attentions elsewhere, others followed the example he had set, and by the end of the 19th century were turning out portable engines as fast as possible. The countryside, eager and demanding, had absorbed no less than thirty five thousand engines by that time. While for Howden it meant oblivion, it was the age of the portable engine that saw many of the great names of agriculture come to the fore, men like Robert Ransome, founder of the famous 'Ransomes of Ipswich', agriculture and steam exponents who did so much for farming in general.

Robert Ransome, a school master's son born in 1753, had gone into an ironmonger's business straight from school. After an apprenticeship period he opened his own business and also acquired an iron foundry. This fired his imagination, and experimental work led to him being granted a patent for tempering plough shares then made from cast iron.

The firm's work was not only confined to agriculture, for as well as building and installing Ipswich's gasometer and the pipes necessary for the town's supply, they were also responsible for the Second Stoke Bridge. In 1830 the founder died, but the company had acquired experience in all aspects of foundry work, and an expansion programme saw them using that knowledge in the field of agriculture, and of course into steam.

By 1841 Ransome's first portable engine was exhibited at the Royal Show at Liverpool. It was, although few attending realised it, an historic moment, for the company had ideas that if they had been taken up would have led to a much earlier birth of the traction engine as we know it today.

At the 1842 Royal Show at Bristol, last year's portable was this year's self propelled. For the farmer it was a completely new innovation. Not only had a sprocket been fitted to the engine shaft driving a larger sprocket on one of the rear wheels, hence giving self propulsion, but the chassis had been lengthened to accommodate a small threshing machine. Here was the complete agricultural investment and reports of its performance were enthusiastic. The power developed exceeded 5 h.p. with an hourly intake of 36 gallons of water and 50 lbs of coal. Despite the fact that a horse was needed between the shafts for steerage, it earned the top award of £300. Unfortunately the Ransome attempt at giving the farmer a package deal was a failure. Perhaps it was the fact that the purchaser had to take the threshing drum, perhaps it was the diehard attitude resistant to change, content with the portable that had served until then. Whatever the reason it did not find favour in agricultural circles and the machine was dismantled.

Despite the setback that this new innovation suffered, manufacturers are creatures of perseverance and the design and improvement of the portable continued. Until 1847 a vertical cylinder was favoured, but now they began to move towards the horizontal and within a year the 'modern design' could be said to have arrived. The builders, Clayton and Shuttleworth, used a horizontal unjacketed but wood lagged cylinder on the fire box top, driving forward to a bent crankshaft carried in cast brackets on the boiler top. It was available in three sizes, 4 h.p.; 5 h.p.; and 6 h.p. Now it seemed that steam had been successfully harnessed for the agriculturist, and many who worked these engines must have doubted if further improvements were possible. The drawback of the portable however, was the need for horses to move it from place to place. This put an enormous strain on the animals, especially during the winter months, and many died from the exertion.

Ransome's saw the problem and again came forward with a self-propelled 'Farmer's Engine', at the 1849 Royal Show at Leeds. Not only was it transportable under its own power, but also for the first time there was no need to have a horse for steerage. Bevel wheels, operated from the driving position, turned the engine in the direction required. Experts were lavish in their praise. *"Here",* one said, *"is the universal assistant for all farms. It only requires to be known to be appreciated."* Yet they were wrong. Ransome's second attempt to introduce a 'traction engine', as this

An 8 h.p. Foster, made in 1942.

machine undoubtedly was, met with the same failure as their earlier design. In an age opposed to change, the Ransome engine was ignored and another chance for steam to progress in leaps and bounds was lost.

The portable however was now established. There was scarcely a farm that did not have at least one engine employed on such diverse tasks as driving a threshing drum, saw bench or chaff cutter. In the height of their popularity it was estimated that the east of England alone housed over 60,000. To most farmers they were the answer to a long felt need. Mechanisation was no longer a dream, but a practical proposition. Yet the portable always had that inherent drawback that horses were needed whenever they had to be moved.

Although Ransome's had failed with their 'Farmer's Engine' on a national scale, there were many farmers who, recognising the drawback of the portable, sought an alternative. The man who initially supplied it was a Rochester farmer, Thomas Aveling.

Born in Cambridgeshire in 1824, he was seen during the early years of his life as slow and dim witted. It was a condition brought about by a step-father who ruled him with a rod of iron, making the boy retreat into himself, revealing his true character only when he escaped from his home environment to work for a local farmer. There he showed an innate mechanical ability, and marriage to the neice of his employer established him as a farmer in his own right. It was then that his engineering ability truly came into its own. Implements in those days were primitive, and breakdowns were a regular occurrence. In the beginning Thomas Aveling satisfied himself by repairing appliances for his immediate neighbours. Then he took things a stage further and began adapting existing implements to a more efficient system. From this it was a short step to establishing his own engineering works at Rochester in 1850.

From the beginning he had been impressed with the idea of applying mechanical power to the field of agriculture, so when the portable engine established itself, Aveling welcomed the advances it brought, yet, as he said, "It is an insult to mechanical science to see half a dozen horses drag along a steam engine, and the sight of six sailing vessels towing along a steamer would certainly not be more ridiculous". So it was no surprise to those who knew him when Aveling adapted a Clayton and Shuttleworth portable engine to self-propulsion, by means of a long driving chain between the crank shaft and rear axle. True, the engine still needed a horse for steerage and this, in the eyes of many was its drawback, for the horse between the shafts, with no actual heavy work to do, developed a tendency towards laziness that often made it useless when put to work pulling loads or implements.

However, Aveling was not finished. The conversion of the portable was but one of the many great ideas he had, and in 1859 he took out his first traction engine patent. At that time he did not have the manufacturing resources capable of building traction engines, so the Lincoln firm of Clayton and Shuttleworth did it for him. The first engine made by them to Aveling's design was exhibited at the Royal Agricultural Society of England Show at Canterbury in 1860, being advertised as an "8 h.p. Patent Steam Locomotive Engine". In appearance it differed little from that first converted portable, even the shafts were still retained, for steering was done by means of a single front wheel, mounted in a fork with a bracket attaching it to the shafts. From the top of the fork a lever extended backwards to be operated by the steersman, who sat with his feet dangling between the shafts.

By 1861 Aveling was in a position to start producing his own engines at Rochester. Soon, at a price of £360, they were finding ready markets in the countryside. Thomas Aveling deserves his place of honour in the story of how steam came to the farm, but other designers were working along side, and adding their ideas to produce what became the ultimate in this form of design. Self propulsion of the portable by means of chain drive was applied by Garretts, Tuxfords, Clayton and Shuttleworth, Savage, Burrell, and others, until it evolved into a definite class of its own, the forerunner of the general purpose Agricultural Traction Engines with which we are familiar today.

These engines were only one aspect of the use of steam as an aid to farming. While Aveling and others were turning their skills to the production of self-propelled engines, men like Thomas Tindall dreamed of the ultimate in steam mechanisation, engines that could actually plough and cultivate the land. At that time the farmer was dependent upon the horse and the bullock. Nothing could be done on the farm without them, yet many realised the inherent disadvantages of these animals. They were expensive to keep, required part of the farm output for their support, and like any animal were subject to illness and death. A mechanical counterpart was their logical successor, and Thomas Tindall believed he had produced just that in 1814. Designed for every form of cultivation, it was a strange looking affair. For propulsion it relied on four legs that successively came into contact with the ground to give it a form of walking action, as the carriage that contained the cultivating attachments was propelled forward. It was however, as its proud owner proclaimed "capable of having ploughs, harrows or even scythes attached to it". Alas for Tindall it was yet another of the failures that dogged early experimenters, but it showed that men were determined to harness steam to the full for the benefit of agriculture.

Designs for cultivation were many and varied. Some tried a digger action to turn over the land, and of course the idea of using a steam engine to tow implements across land was soon tested. Here however, with direct ploughing by steam, weight was the over-riding factor. In fact steam in cultivation was only made possible when men realised that the future lay in the field of 'indirect ploughing', with a stationary engine providing the power for a plough to be drawn across a field on a cable and windlass system.

By 1859, Lord Willoughby d'Eresby, on his estates at Grimsthorpe, Lincolnshire, produced a development of this idea, laying the foundations on which others followed. The engine chosen was of the locomotive type, modified to include detachable drums carrying the cable that was used to haul the plough from one side of the field to the other. At work the engine was run on railway lines laid in the centre of the field, and the cable run out to a capstan in the headlands. A plough advanced and receded from either side and at right angles to the engine, worked by an endless chain from drum to plough, round an anchor cart on the headlands and back again to the engine.

Figures issued at the time spoke of the system doing the work of sixteen normal ploughs using anything up to thirty horses. Only eight men and one horse and cart to fetch water for the engine were now required. Unlike many who, after tests, seemed to fall out of the steam age, Lord Willoughby d'Eresby worked his system in Lincolnshire for the next five years, although when the next development in the search for steam cultivation came, it brought into service the already established portable engine.

The man responsible for making the portable even more of a universal farm engine was a farmer named Hannam, from Burcote, near Abingdon in Berkshire. His ingenuity saw a new piece of equipment enter the story — the windlass. Briefly this consisted of two independent drums mounted on a wheeled base. The drums, belt driven by the farm portable, provided the circular motion to let out and take in a cable to a plough in a similar manner to that used on the Grimsthorpe Estates. Now with just one piece of extra equipment and the cable, the farmer with his portable had a steam ploughing system.

Hannam's idea, which was used successfully until the introduction of

92

Built to original works plans and mouldings of 1860, this 1975 Savage chain driven engine shows one of the first attempts to give agriculture a self-propelled engine, in this case with a separate steersman at the front.

the great pairs of ploughing engines, became known as the 'roundabout' system. In use, a free running rope of iron wire was taken round the field to be ploughed, through guide pulleys positioned in the corners. The engine was coupled by a belt to the windlass, which in turn was linked to the rope. This could now be run from the windlass, round the field and back, thus making a complete circuit. The plough, which in those days was a normal horse drawn implement, was incorporated at one edge. There were drawbacks. Several farm workers were needed to move the pulley after each crossing as the area still to be ploughed narrowed, but it had the advantage of providing any farmer who owned a portable engine with a ploughing system at little extra cost.

The 1850's are of course notable for the emergence of the man who became synonymous with every aspect of steam ploughing — John

Fowler. In 1849, at the age of 23, John Fowler, then employed by a Middlesbrough Engineering firm took a holiday in Ireland, an event that was to change his whole life. At the time of his visit Ireland was suffering the aftermath of the terrible potato blight of 1846-47, and Fowler was horrified at the hardships and deprivations of the starving country folk. It was a horror made greater by the sight of vast areas of uncultivated land which he knew, given adequate drainage, could provide the life-giving food that Ireland needed so badly.

By 1850 Fowler was demonstrating a mole plough designed to lay wooden pipes to a depth of two feet in heavy clay soil. In the beginning he used horse or man power to turn the required capstan. By 1854 he had a drain plough with anchors and windlass driven by the now familiar portable steam engine. Then inevitably, after his success with the mole plough, John Fowler turned his attentions to the general field of steam cultivation.

There was at that time, a prize offered by the Royal Agricultural Society of England, for the first truly proven system of steam cultivation that could be used on land at a more economical and efficient rate than the old time honoured methods. Fowler determined to win that prize. He began by experimenting with the 'roundabout' system, but it was a chance meeting with a Scottish farmer, John Greig, then living in Essex that saw Fowler make the breakthrough. So impressed was Fowler with Greig's grasp of the difficulties involved in steam ploughing, that he offered him a job as technical advisor. It was a wise choice, for the two men working together designed a plough specially for steam cultivation. Up to then the old horse drawn types had of necessity to be used, with the inherent disadvantage that the shares faced in only one direction. In 1856 Fowler and Greig took out a patent for a balance plough consisting of two sets of four shares, one set left handed and the other right handed. In use one set of shares was dropped to the ground and the other rode in the air. On completion of the pull across the field, all that was needed was to drop the airborne shares, the others were raised and the plough was ready for the return crossing without having to be manhandled into a reverse position.

The plough was to be the basis of all future ploughs used during the years when steam had the ascendency in the field of cultivation.

Fowler's system was similar to that already in use, with a double drum windlass being connected by a belt to a portable engine. A cable from one drum was run out round a pulley, across to another anchor pulley, and thus in a straight line down the field to be worked, to the plough. Fowler,

94

One of a pair of Fowler ploughing engines used to haul the great anti balance ploughs. No. 14381, a 16 h.p., 26 ton, model BB, built in 1917, was part of the intense campaign to improve our food production in the face of the U Boat blockade.

of course, had the balance plough and that added 100 per cent improvement, bringing him in 1858 the £500 R.A.S.E. prize given after a demonstration at their show at Chester.

The following year Fowler really proved his supremacy in the field of steam cultivation, for at the Royal Show his three and four furrow balance ploughs proved their worth over every other type entered.

In the beginning outside manufacturers carried out Fowler's constructional work, but by 1859 he knew that if steam ploughing was to succeed, engines of greater power were needed. So, in 1860, with this purpose in mind he opened his own works at Hunslet, Leeds.

Even before this John Fowler had tried ploughing with two engines, one on either side of the field, each alternatively drawing the balance plough across the land. The advantage lay in doing away with the need for anchor pulleys, but the disadvantage in the minds of many farmers was the need for two engines. It was however a two engined set that had the distinction of being the first 'Fowler Plougher' to leave the Hunslet works. In these the cable drum was slung horizontally below the boiler as an inte-

95

gral part of the engine, and with the windlass done away with, the ploughing engines were a self-contained entirety. For those who still wanted 'roundabout' working, Fowler made and sold an engine with double drums fitted. One was used to haul in the cable as the other let it out, with the positions being reversed for the return haul.

By 1863 Fowler and steam ploughing were firmly established in the rural landscape, and while the ploughing engines had not attained the standard that we see today at meetings and rallies, the basic design was there. Alas, Fowler himself was not to live to see their development, for he died in 1864 after being thrown from his horse when he was recuperating from the effects of overwork. Not that the death of Fowler brought an end to progress at the Hunslet works. For as well as being a brilliant engineer, John Fowler had the ability to select and establish a team around him capable of the same dedication that he had given to the task. So the works were reorganised bringing David Greig, the brains behind the balance plough, into the business as a partner with direct responsibility for technical design. The commercial work was undertaken by John Fowler's

A 1918 Garrett Suffolk Punch tractor No. 33180, designed to meet the threat of the tractor for direct ploughing.

brother, Robert. At the time of the change, the 'Steam Plough Works' employed 600 people with one machine rolling out of the factory gates each week.

It was shortly after the death of the founder that the engines which were the predecessors of those we see today were introduced. These were the Fowler horizontal shaft engines. The drive to the winding drum fitted below the boiler was by means of a horizontal shaft and bevel gearing meshing, with the crankshaft at one end and the drive drum shaft at the other. This is the feature that became standard on the great cultivators, and during their entire working life no better system was discovered.

Although Fowlers gained ascendency in the field of cultivation, other makers did not surrender entirely to him. Famous names in steam, Aveling and Porter, Ruxford and Sons, Savage Bros, Marshal and Sons, J.H. McLaren, and others, all built engines for the purpose of steam ploughing, but the credit for the breakthrough which made it a sound commercial proposition must be given to John Fowler and those who carried on his work.

By 1902 a really good farm tractor had appeared and although even in 1910, when steam was tested against the tractor, it more than held its own, but the writing was on the wall and it was only a matter of time before a new form of motive power came to serve the farmer.

However, steam did not surrender without putting up a fight. In 1917, the year that the British Government placed a mass order for American tractors in the fight for survival against the German 'U Boat' blockade, Richard Garrett and Co. of Leiston, put on the market their 'Suffolk Punch' steam tractor intended to draw the plough or other implement behind it as the tractor did. Weight, however, was still against steam and this, and the Mann & Co direct steam tractor both failed to halt the progress of the internal combustion engine.

By 1930 the tractor had almost completely taken over from steam. It was said, by a prominent agriculturist of the period, that an internal combustion caterpillar tractor with a six furrow plough could turn over ten acres in one day. True, the steam ploughers could work twelve acres, but they not only needed the two giant engines and plough, but also at least five men to work them, as against the single man who drove the tractor.

7 · Farm Transport

In the days before steam arrived on the farming scene it was the horse that provided most of the motive power. Indeed, even after steam had taken over many of the farming tasks, the horse was still a necessity, ensuring the life blood of many rural communities, for a village in those days was a self-contained unit. One of quite moderate size, say two hundred to five hundred occupants, would have several tailors, a miller, blacksmith, carpenter, saddler and wheelwright as well as those concerned with the selling of food. Supplies from the towns and cities were essential for most forms of trade. Hence no community was too small to have a carter, the man who made his way into the outside world once or twice a week to return with the needs that could not be fulfilled locally.

It can be said that in the evolution of agriculture, one of the most important items was the humble cart and waggon. It formed the backbone of every aspect of farming, and as well as supplying the needs of the village carrier, it also served in towns and cities for the carriage of goods and people, but it was the cart and waggon designed for the innumerable tasks around the farm which today brings forth a recognition of the innate skills of those craftsmen on whom the demands of the village fell.

Alas, the world of the internal combustion engine had no place for the steady pace of the horse age, and its replacement by another form of transport was inevitable. Today we can only marvel at the endurance of these masterpieces of skill. Although progress may have rendered them obsolete, they in their turn have shown up modern methods and materials by proudly living on into our own period, with a longevity that modern production techniques dare not contemplate, and through those that have survived, be it in rural museums, the hands of private owners, or just left to slowly rot in some hedgerow or derelict barn, their natural elegance derived from both function and design shines through. Such creations do even more, for they stand symbolic of the craftsmen who held a position of envy in any rural community, the wheelwrights. For the English farm waggon, at the peak of its era, constituted one of the best examples of what we mean when we speak of the finest in craftsmanship. Yet often the work of the wheelwright went even further than that; he could also be the village carpenter and joiner. The job of the joiner was to carry out the construction of woodwork in buildings, while the carpenter's trade dealt with waggons, carts and carriages, so the country craftsman who carried the title of carpenter and wheelwright was the man who saw the job of waggon and cart building through to finality, with of course help from the blacksmith.

Often it was this dema cation of trades which saw the work of waggon building as true teamwork in every respect. The carpenter would be responsible for the body, the wheelwright for the wheels and then the blacksmith for making the forge work. Finally came the painter, often grinding and mixing his own colours, to give the waggon its final artistic touch. An example of the durability of the painter's work is that when a waggon is seen today the date it proudly bears does not indicate when it was made, but often when it was last repainted!

Quality was the keynote in construction. The waggon was built by men who went into the woodlands, selected, and often felled the timber themselves. Then when they got it back to the yard it would be cut with a giant cross cut over the saw pit.

If an apprentice was engaged in the task, it was he who would work in the bottom of the pit and receive all the sawdust. Anyone lucky enough to be apprenticed to such a tradesman was envied by those less fortunate, for it meant that his parents were wealthy enough to be able to raise the bond, usually between £5 and £10, and were prepared to keep him in clothing for the five to seven years during which he was 'serving his time'. The hours were long and hard but the job was held in esteem as few village lads had the chance to learn the craft.

Like so many other trades, those of wheelwright or carpenter were handed down from father to son to be carried on in the time honoured way, serving the needs of the community.

When the timber had been cut it had to be seasoned, a long process with usually one year being given for every inch of thickness. It was care like this that made the waggon and cart the durable vehicles they were. More than this, it was the fact that generations of experience produced men who knew wood in all its moods and knew which type would serve a particular need best.

Elm was invariably used for the nave or hub because only this would stand up to the making of the holes for the spokes, often numbering fourteen, and the tapered hole for the axle. Only elm could withstand the strain after the spokes had been knocked home, of years of use on tracks that passed for roads in that period. Again elm came into its own for the panelling and flooring, with ash and oak serving for the framework of the body. Oak was also used for the spokes of the wheel and was cleft, or split, as opposed to sawn, as this method showed faults that would have been invisible until the spoke cracked under the strain as the cooling tyre contracted.

More than a knowledge of materials was needed however to build a waggon which would serve a farmer over the decades, and nowhere is the inherent skill of the builder shown better than in the wheelwright's work on the wheels. For while his ability to provide a wheel that ran true was the basis of the cart or waggon, it is the 'dished' shape of the wheel that experts admire so much today, for this is one of the strongest constructional methods possible.

However, those early woodworkers had no such knowledge. All they knew, through the hard school of experience, was that as the horse moved in the shafts, the swaying of the cart gave a side to side motion to the wheels. Straight wheels would have broken apart under the strain, and the 'dished' shape was evolved.

Another consideration when it came to using the dish shape was that it kept the top of the wheel away from the side of the vehicle. To this end

A Lincolnshire waggon, now neglected but still showing the grace of the early farm transport.

axle arms were made with a definite dip downwards to give positive clearance. Obviously, if the straight wheel was used and mounted on such an axle, then the spokes would not be vertically opposed to the ground — rather they would slope inwards and the wheel would never support the weight, even at rest.

So although these old wheels may have looked simple affairs, they reflect the true ability of those craftsmen of yesteryear, solving all the problems of sideways thrust, the need to clear the top of the cart and, of course, to keep the spokes vertical to the ground.

While it is the farm waggon that usually catches the eye, it was the humble two wheel cart that was the true carrier on the farm. A cart was of course less costly and could be worked by one horse. It is interesting to compare the pulling ability of the horse when it was used with the cart as opposed to the waggon. A horse harnessed to a cart could, on a good road surface, exert a pull of just over 50 lbs to move one ton. This compared well with over 65 lbs needed for a similar effort per horse in a waggon. On agricultural land the difference was even more in favour of the cart. Here the figure was 200 lbs as opposed to the 295 lbs for the same load in a two horse waggon.

These figures were illustrated at a demonstration at Grantham, Lincolnshire in 1850 when 5 horses in five carts were matched against ten horses in 5 waggons. The carts beat the waggons by 2 loads in a day's work. In fact an old agricultural handbook, taking into account that a cart with its single horse could carry up to 22 cwts, suggested that three were adequate for a holding of one hundred acres. A waggon only came into its own if that acreage was doubled, when one waggon and four carts would be needed. For the larger farms, those of five hundred acres, two waggons and eight carts were the norm.

In some parts of the country carts were used exclusively due to the nature of the land. In such areas, with narrow winding lanes, the waggon, requiring as it does a greater amount of room for turning, was impracticable, with the same applying to hilly places where the cart reigned supreme. Another aspect in favour of the cart was that although the waggon could convey a very heavy load, it put no stress on the horse's back, and a horse draws a load all the better and the steadier by bearing a proportion of the load on its back.

Among carts, the tilt cart was the most important vehicle of transport on the farm and was used for nine-tenths of all cartage operations. It carried manure, conveyed stone and other materials for drainage and of

A typical Lincolnshire cart of the type common on all farms during the age of the horse.

course brought home the produce from the fields and took it to market.

Even on something as simple as the cart, the experience gained over the years by those concerned in its manufacture was applied, so the user got the maximum benefit. Nowhere is this better illustrated than in the floor. Here the boards were deliberately placed parallel to the sides; in the technical jargon of the carpenter they were 'long boarded'. This was done for a very practical reason. The task of raking out the contents of the cart, even in the tipped position, would have been very frustrating if the boards had been laid sideways. In the parallel position, especially as over the years the boards showed some degree of warping, raking out was a simple operation as the rake itself came down the lengths of the joints.

While the tilt cart was an admirable form of transportation during most of the year, the need for extra carrying capacity came at harvest time, when the farmer, trying to take advantage of the weather, needed the maximum amount of space. To this end the cart could be fitted with frameworks, called harvest ladders, to the front and rear, hence enabling even more sheaves of corn to be carried.

In the East Midlands at times such as this, the Hermaphrodite came into its own. This was basically a two-wheeled cart that could be coupled to a waggon type forecarriage, and with a framework supporting a large horizontal ladder extending forward over the shafts it was to all intents and purposes a farm waggon.

Every part of the cart was designed with a view to practicality. Even the intricate shapes which gave both carts and waggons a character of their own, resulted from consideration for the horse. For although decorative works of art when the painter had finished, their prime purpose was to lighten the total weight without weakening the basic structure.

Compared with the cart the ancestry of which can be traced back towards the end of the 11th century when they were hauled by oxen, the waggon was a recent innovation. It is said they were introduced into the country in the 16th century by Dutchmen who came to drain the fens, and many East Anglian waggons have retained some Dutch features.

The waggon can be divided into two main parts, the under-carriage and the body. Yet while the under-carriage tended to be traditional throughout the country, the bodies varied according to the geographic location.

The basis of the under-carriage was of course the front and rear axles which were linked together by a coupling pole braced to the rear axle. Above the brace was a bolster that supported the waggon body. A similar system was impossible with the front axle for this gave the waggon its lock or turning ability.

To this end 'hounds' were fitted to the axle, carrying a 'sway bar' on which the pole rested. With the front end of the coupling pole resting on the axle and secured with a king pin that came through from the bolster above it, the system was simple but effective.

Such however was the perfection of the craftsmen who produced the waggons for the farmer that any component, be it of wood or iron, could be removed for replacement or repair. Nor was any part of the waggon secured by glue. In earlier waggons the strake or iron tyre was made in sections and nailed to the wheel. Strakes were in segments corresponding in number to the felloes, each strake overlapping the felloe joints and secured by five nails at each end. The nails, needless to say, were made in the village blacksmith's shop and had flat shanks and square ends. Later on, the wheels were fitted with 'hoops' — continuous metal 'tyres.' In either case, however, they were made in a special furnace and put on hot, so that in cooling they contracted onto the wheel and bound them as tight as if in a permanent vice. The circumference measurement was vital, and this was taken on the flat strip by using a 'traveller', a disc on a handle that resembled in part the surveyor's wheel that is used today for measuring purposes.

When the strip had been cut it was put through rollers to produce the required hoop. Exact temperature of the forge fire was essential if the end of the hoop was to shut as required. Then when the ends had been joined, the hoop was put back in the furnace to expand so that it could be dropped over the wheel, which had been screwed down to an iron-tyring platform. Tongs were used to carry the hoop to the wheel and while it was knocked home with a sledge hammer, men stood by with watering cans to extinguish any wood that ignited under the heat. As it cooled and the metal contracted so it forced home felloes and spokes in a manner that would have been impossible using any other method.

In the earliest waggons all four wheels were large both in diameter and tread, but the later trend was for a reduction in size, particularly in the case of the front wheels, as attempts were made to reduce the distance needed to turn a waggon. It was true to say that the first designs took nearly half a field when it came to turning, but as these old craftsmen gained experience they began to build a waggon that could turn in its own length.

In fact it was this inability to turn in a short distance that condemned many waggons to the scrap heap when the tractor was introduced. Where they were still in use, with tractor as opposed to horse power, those with limited lock proved useless with this new form of motive power.

Similar care, of course, was given to all the other stages in the building of the farm waggon. An example of the skill that generations of country carpenters acquired is shown by the bevelling or champering that is to be seen on most examples preserved today. It was estimated, by those connected with the craft that up to one-eighth of the original weight in wood could safely be eliminated, while still avoiding the parts that needed maximum strength.

This champering was itself traditional and peculiar to a region. Some areas, such as Suffolk and Kent, produced waggons with only the minimum of champering. Others, counties like Herefordshire, Somerset and Dorset went to the extreme and devotees of the craft turned the final

The Hermaphrodite cart; a two-wheeled cart which could be coupled to a waggon type forecarriage and used during the harvest period.

appearance of their handiwork into a labour of love. The finished product also owed much to the skill and artistry of the painter, and in many respects his designs on the head and tail boards of many of the waggons can be compared with the work of the canal artists who so colourfully decorated the narrow boats.

Again there was variety according to the region or county. In Somerset and Dorset the artist followed the example of the carpenter who devoted love and care to shaping and finishing the wood, and their motifs and lettering appeared in brilliant colours. Other regions, however, showed a total absence of motif, satisfied with the simple charm of pure craftsmenship.

For the paint work, red was almost universally selected for the wheels and under frame, while the body carried whichever colour had been used in that region for as long as man could remember.

Lincolnshire would finish its carts in either all red or a red underframe with a prussian blue body; but the difference was greater than just colour. The spindle-sided vehicle favoured in these parts was a difficult waggon to work with. The body was pitched well above the axles, and with its high sides and wheels measuring anything up to 60 inches in diameter, it could make work in the harvest fields demanding indeed.

On the opposite side of the River Humber, in the East and West Ridings of Yorkshire, the contrast could not have been more extreme, as the charm of the spindle-sided waggon gave way to the rough utility of planked sides. Again, like their Lincolnshire neighbours, the waggon makers of the county favoured red for the body colour with brown as the alternative colour in certain parts of the region. These are but two examples, for wherever one travelled, regions that used the waggon as part of their farm transport had their own individual styles and colours. This was inevitable for they were the work of individuals, men who, through the years, had learned the requirements of the farmers, and who never travelled outside the community, and hence could not be impressed by the work of others. Yet while one may be inclined to believe that one form of design was better than another, a Somerset waggon would have looked as out of place in the Wolds of Lincolnshire as would a Devon waggon in Wiltshire. Somehow the waggon, the product of a region, blended with that region and its landscape, and was at home only in the area where it truly belonged.

Building the waggon was one thing, but making it safe in use, particularly in the negotiation of hills, was another. Again the answers to this problem were varied and ingenious. There was the 'Dog stick', a wooden stick with a forked metal end that was fitted to an axle tree. In use, when the cart was negotiating hills, the 'dog stick' was allowed to trail on the ground. Then if the waggon stopped, the 'forked end' pushed itself into the ground and prevented the vehicle moving backwards of its own accord. Another strangely named device used for obtaining a braking effect was the 'Drug bat' or 'Skid pan'. This was a cast metal device designed in the form of a wedge shaped shoe, and when it was placed under a back wheel it would skid and hence lessen the speed of descent of the waggon.

Another similar idea, the 'roller scotch' was a cylindrical wooden 'brake' that rolled to the rear of the back wheel. Then if the waggon stopped or began to run backwards when going uphill the 'roller scotch' scotched the motion. The 'locking chain' was the final braking safeguard.

This was simply two pieces of chain, a long length passed round the rim of the wheel and secured to the shorter piece that had a ring and dog-hook on it.

The waggon is truly representative of the ingenuity of those craftsmen of yesteryear and symbolic of skills that were learned and executed with pride. It also illustrates with many of their regional differences, the conditions that existed when they were part of the farming scene.

In the days before Macadam, the by-roads, lanes and farm tracks were often nothing more than a morass of mud, and cart ruts became so deep as to become permanent. The result was to impose a degree of standardisation on the wheel tracks for all forms of transport in a region. They became, in effect, similar to the rails of a railway, deepening with the passing of each winter, and baking through the months of summer. So similarity had to be the rule for all carts and waggons that used them, and hence the track that was used originally was maintained by generation after generation of cart and waggon builders until it became representative of the region.

To the farmer the waggon was truly symbolic of life itself. For although it was primarily designed for the harvest field, it was the waggon that was pressed into use on occasions like the school outing, or the wedding of the farmer's daughter, and the funeral of the farmer himself when the waggon would be cleaned, and decked with floral tributes.

There is another aspect of the carpenter's craft that has now been forgotten but occupied a good deal of time in the age of the horse and traction engine. This was in the making of the wheel barrow which served a multitude of duties around the farm. These were the work of local craftsmen and, like the waggon, they evolved with differences in design according to the district.

8 · The Tractor

In 1910 steam must have been secure. True, the internal combustion engined tractor had appeared on the rural scene, but when put to the test, working side by side with steam at trials held that year by the Royal Agricultural Society, it was the latter that got the vote of confidence. There were far sighted men in the world of steam who gave a shudder of fear as they gazed at these new arrivals. Casting their minds back to the first primitive offerings in steam, these tractors must have stood symbolic of progress from which there was to be no escape.

There was, indeed, cause for fear. Progress in any field, once man has made the initial breakthrough, is inevitable as development takes a logical course. The first tractors were far from ideal, but they were the foundations on which men were to build to give us the tractor as we know it today. Yet it was not the tractor that brought the first competitor, but rather the stationary engine. In many ways comparable with the portable steam engine (for neither were capable of independent movement), the stationary engine began the revolution that brought total success for the internal combustion engine, the brain child of Herbert Stuart, a Yorkshire engineer, who with Charles Binney produced a stationary engine to run on either paraffin or lamp oil, in 1890. The manufacture was taken up by the Grantham based company of Hornsby and Sons, and so successful did they become that by 1894, the Royal Agricultural Society had organised tests to discover the most suitable for farm use.

The method of starting was clumsy and time consuming, in that the fuel had to be pre-heated by a blow lamp, but once started they would run reliably for as long as the farmer needed their power.

Naturally they were seen as the basis of a form of general farm motive power and to this end Hornsby and Sons, as well as Ransome produced versions fitted with wheels and capable of being moved about the farm. However, they were too large and heavy, particularly when taken onto soft land. In fact many makers were satisfied that power and performance would never be achieved without involving excessive weight in the construction.

There were others however, who persevered, and while credit for development of the first tractor is given to America, there was much work done to perfect it on this side of the Atlantic. Men like Dan Albone, a Bedfordshire engineer with a farming background, foresaw the future and by 1902, when still in his early twenties, he produced his first farm tractor. That was also the year when John Froelich of Iowa introduced what was the forerunner of the now famed John Deere line of tractors. In fact it was Dan Albone who gave this country its first tractor industry with the establishment of *Ivel Agricultural Motors Ltd,* which made over 900 tractors before it collapsed during the economic depression of the early 1920s.

John Froelich's *Waterloo Tractor Company* was another that failed to last through this bleak period in farming life, but it had already been taken over by John Deere, whose tractors are world famous today.

The *Ivel* range of tractors were of a tricycle appearance, with two large drive wheels at the back and a smaller steering wheel at the front. Weather protection for the user was of course non existent in those pioneer days, but the tractor, which developed 24 h.p. well deserved the Silver Medal it won at the Royal Show two years after making its debut.

Ransome, Simms and Jefferies, the already established implement makers were another concern which turned its attentions to giving the farmer this form of motive power, but unlike Albone's three wheel approach, their design, with a 20 h.p. four cylinder engine, was built in a similar style to the cars of that age.

In many ways it claimed to be the universal power source for the farm, able to plough, drive threshing machines and saws, as well as pull binders, reapers and mowers. Alas, there was a general lack of interest, due mainly to the high cost of over four hundred pounds, and this was the main reason why the farming community turned its back on the Ransome offering. However, the claims put forward for the tractor were such that they could not help but excite the farmers' interest. In any rural community, the idea of mowing three and a half acres of grass in one hour was more than appealing. So it may have seemed to many that the Agricultural Trials held in 1910 would see the tractor take a further step forward, as four internal combustion engined tractors assembled to prove their worth against three steam engines.

Ivel Agricultural Motors entered two tractors and the *Saunderson and Mills Company,* also experimenting in the field, entered a 25-30 h.p. and a larger 45-50 h.p. For months the judges held the farming community in suspense while they deliberated on what they had seen. Then when their final judgement came it was a form of condemnation of the tractor as it was then. The requirements, the judges said, were for a tractor that would be light and yet capable of doing the field work which was then the pro-

An extension which could be fitted to the tractor wheels to aid adhesion and grip on heavy land.

vince of the steam contractors. None of the tractors which had performed in front of them reached the required specifications.

With makers still quoting a figure in excess of £400, price was the main deterrent. The farmers also disliked the idea of holding stocks of inflamable petrol. Petrol was necessary to start these early tractors, and to warm them up before the switch-over to paraffin, and although today it may seem a small thing, it was one of the factors that decided many farmers against the tractor.

The breakthrough, when it came, was not the result of any changed agricultural thinking, but rather the war clouds that darkened the skies over Europe, resulting in the outbreak of World War I in 1914.

However the tractor had not stood still in the years between the 1910 trials and the beginning of the war. Ivel, Saunderson, and others had extended the range of tractors available to the British farmer, but it was war that gave them their final impetus. The threat of a German 'U' boat blockade of our coasts meant that we had, of necessity, to produce more food and put more of our countryside under the plough. Farm workers were among those sent out to fight the war in Europe, and farm horses also found themselves serving the needs of the army. So those left to work on the farms found that whereas demand had increased the means to supply that demand had decreased.

It was in America that the tractor really began to dominate the farming scene, due to the activities of Henry Ford. He was already heavily committed to producing his 'T' model car, as well as others, but that did not stop him turning his attention to the tractor. As a farmer's son, it was the possibility of tractor production that occupied his mind before he turned to the motor car. In many ways his first offering to the farming community provided the best of both worlds, being in the form of a conversion kit that could change the standard Ford car into a tractor.

It was fortunate for Britain that Henry Ford was involved in the mass production of tractors, because, by 1917 the 'U' boat blockade had been so successful that our food supplies were only sufficient for a few weeks, and despite valiant attempts by the home tractor industry, an influx, greater than it could supply was vital. An urgent order was sent to Henry Ford for 5,000 American tractors. It was an action that not only served the immediate purpose of providing additional food supplies, but led to an attitude of mind among those handling them, which made them more receptive to the idea of future agricultural mechanisation.

The 'Model F' Fordson tractor, our 1917 saviour, was in advance of many of the early experimental tractors used in England. They were fitted with steel wheels, had worm and wheel final drive, but above all showed the concept that Ford always visualised in his search for the perfect farm vehicle. Frames, cross members, chassis, all these had been dispensed with and basically what they provided was an integral engine and gear box to which were fitted the two sets of wheels. Ford had cut the weight to a quarter of what his competitors achieved but retained quality workmanship with first class materials. Add a drastic reduction in price as compared with other tractors being offered to the farmer, and the reason for the Fordson success story is easy to see.

Henry Ford, of course, was not the only American tractor manufacturer, even if he did turn out to be the most prolific, claiming almost three quarters of the market. The Case Company, now affiliated to the British

111

tractor Company of David Brown, built their first tractor in 1892, pre-dating even John Froelich. This however had a traction engine appearance and was plagued by ignition trouble. In fact, in those pioneer years many American builders sought to use the internal combustion engine in a tractor that looked in fact more like a traction engine than the tractor we know today. The evolution, now it had got under way saw the old ideas vanish and a new breed of tractor come into being. However, in general, America turned to the heavier, larger tractor, the type demanded by their own vast acres. Although England needed a smaller, lighter tractor, some of the American models found their way across the Atlantic and gave excellent service. The International Harvester Company exported their Mogul and Titan tractors into England from 1914 onwards, and many of these, like the later Forsdon tractors were hired out by the Government. The Mogul was a 16 h.p., four-wheel model, while its brother, the Titan, was a 25 h.p. tractor of which over 3,000 found their way into Britain by the time production came to an end in 1920.

Possibly International Harvester's most important contribution in the

An International Mogul 10/20 built by the International Harvester Company, Chicago. This had a single cylinder engine with 8½″ bore, giving a normal r.p.m. of 400.

field of tractor development came in 1918, with the introduction of its power take-off mechanism. Prior to this, when implements being towed behind the tractor needed power to operate they had to take it from their own ground wheels, (in the same way as horsedrawn implements) with the inevitable trouble caused by slipping. Now, through a special shaft fitted to the tractor, power was provided direct to the implement from the tractor engine.

Harry Ferguson also made a large contribution in relation to the mounting of implements on the tractor. Prior to the arrival of the tractor, in the days when the horse was commonly used, the 'pull' on the implement was applied fairly high, from the neck and shoulders. The tractor was used at first with the same implements as used with the horse, so the pull still had to be applied from a fairly high position. Harry Ferguson's idea saw the implement become an integral part of the tractor with a 'three point linkage' for attachment. Two bottom links drew the implement along, with an upper link applying force from the implement back to the tractor, hence aiding traction of the back wheels of the tractor.

By 1935 Ferguson had perfected the system with a complementary hydraulic system suitable for a wide range of farm implements. It was this, in terms of practical farming that signalled the true arrival of the tractor in agriculture. Previously the tractor and implement were separate items of farm equipment. Now they were one, with the tractor driver able to position and operate his implement from the driving cab.

In the beginning tractors were made to Harry Ferguson's designs in a factory in Huddersfield, but this ceased when he verbally agreed that Ford should take over the constructional side. However there was disagreement between Ferguson and Henry Ford Junior, which led to a law suit and the award of over three million pounds in damages to Harry Ferguson for past royalties. On top of this Ford was stopped from building the Ferguson tractor, which Harry Ferguson himself undertook from a factory in Coventry until 1953 when he amalgamated with the Canadian, Massey Harris Ltd., to bring into being Massey Ferguson products.

All this, however, was still in the future when World War I came to an end, and agriculture and the tractor and implement manufacturers tried to adjust to peacetime. By this period there were few farmers who did not accept that the tractor was here to stay, but the trouble, for the farmer trying to decide on a choice, was the large number of different models on the market. In 1918 there were one hundred and forty different makes with more to come during the next year.

What the farmer needed was organised trials, and it was the Society of Motor Manufacturers and Traders which gave him help in that way in 1919 with trials held near Lincoln before independent judges. At the trials the organisers did everything they could to see the farmer got the information he needed. In addition to agricultural judges, a consulting engineer was present to check over the mechanical aspects of tractors which had to show their ability at ploughing, threshing, cultivating and hauling. To make the test really representative, the organisers tried to get one example of every type of tractor being sold in England.

Many of the names of companies that took part have alas, sunk into oblivion as far as tractors are concerned, names like Alldays and Onions; Avery; D.L. Motor Company; Omnitractor; as well as Martin's Cultivator Company, and the pick tractor, both from Stamford in Lincolnshire. In 1919 it was firms like these which were trying to give farmers the tractor that would fill all his agricultural needs.

However, the trials proved that with tractor design as it was then, there was no universal tractor. The choice had to depend on the type of land and the work needed from the tractor. The importance of tractors being seen working under typical agricultural conditions could not be over stressed in those early days of their life. To this end the Royal Agricultural Society held their own tests, again near Lincoln, in 1920. There were seven classes. Five for the internal combustion-engined tractor, and two for steam, one for direct steam cultivation and the other for the double engine cable steam plough. It is interesting to reflect that there was a similar double engine ploughing class for the internal combustion engine and Fowler in fact made some twin ploughing units using this new form of power. In this field the internal combustion engine could never hope to succeed. The farmer was looking for a light tractor for direct ploughing and few would be interested in buying a system that had two engines, neither were the contractors, the men who in the days of steam had invested their capital in ploughing sets to hire out to the farmer. As the tractor began to prove itself more and more, so farmers bought these single units to do away with the need of bringing in outside labour for farm cultivation.

The classes concerned with direct cultivation were chosen according to the horse power of the engine and everything was done to ensure that each tractor worked on both light and heavy soil, as well as on hilly land where stopping tests, both going up and coming down, were carried out.

The importance of the trials to the manufacturers was shown by the interest created abroad. Tractors from America and Switzerland were there, as well as entries from Scotland and Ireland. Of the thirty eight tractors that actually worked the trials, only two had to retire, a fact that brought favourable comment from the judges, who pointed out that many would have to operate in remote isolated areas, well away from workshops, so reliability was essential.

In the year 1920 Harry Ferguson first brought out his 'three point linkage', and its development ran parallel with tractor development in general, as makers sought to improve on the already impressive ability shown at the trial of that year.

Tractors, of course, run on tracks as well as wheels, and both types have their use. On heavy, wet soils it is the track vehicle that comes into its own. A wheeled tractor on such land could encounter wheel slip, causing physical damage to the soil and making the job difficult for the driver. The track vehicle on the other hand, distributes the weight over the track area and can work in conditions that would possibly bring a conventional tractor to a halt. The track vehicle is slower than its wheeled counterpart, less versatile and more expensive. On the credit side however, it does enable field work to start early in the spring and carry on until late in the autumn.

However, one of the more interesting aspects in the story of the tractor is found in another example of a lost opportunity. As early as 1906 Richard Hornsbys had produced just such a vehicle and on its tests it proved itself to such an extent that the factory prepared for an influx of orders. Alas, they failed to come. Agriculture, it seemed, could see no value in this addition to the farm tractor range. So the company, happy to cut its losses, sold its patent rights to the American Holt Caterpillar Tractor Company. What followed is now history. Not only did the Holt Caterpillar Company persevere with the idea until it made them world famous, but when the value of the track fighting vehicle was finally recognised, it was necessary to import from America before the experiments that led to the first tanks could be carried out.

Since the 1920s tractor progress has in many ways followed parallel with cars. Engines and gear boxes, as well as transmission, have improved out of all recognition. Electric starting has been fitted, and today's tractors are fitted with lights and indicators for use on roads. Cabs have improved, both from the safety aspect and in the reduction of noise, which in the past caused so much driver fatigue. Possibly the biggest difference between the modern tractor and its counterpart of yesteryear is to be found in its tyres.

A Fordson tractor, the 1932 Model N, which illustrates the progression of the Fordson into the modern tractor we recognise today.

The question of tyres and grip presented a big problem for the early tractor designers. In the Fordson Model 'F' which came over to England in the days of World War I, and indeed with other types of that age, adhesion came from slats fitted across the large rear wheels. Rings with spikes were available to put on the wheels when working in really wet land. In turn the smaller front wheels were ridged to give improved steering, but while metal wheels proved adequate on the land, it was when the tractor had to be used on roads that difficulties were encountered.

Under the various Road Traffic Acts, the use of 'slatted' wheels on public roads was prohibited. While it was possible to fit a metal shield round the wheels when the tractor had to be driven on roads, fitting was time consuming.

Solid rubber tyres had been tried, but while they served the purpose as regards road travel, they were ineffective in use on land. In fact the problem stayed with the farmer until 1932, when the Firestone Tyre and Rubber Co. in America introduced a 'slatted' pneumatic tractor tyre. Its value was seen at once by the Allis Chalmers company, another tractor manufacturer, and soon they became standard on all tractors.

Today's modern tyres give the tractor good grip in all conditions, and whenever further adhesion is needed a form of metal cage can be fitted to the wheels and removed with ease for road use.

Tractors possibly lack the appeal of steam, but their evolution was equally important as men sought to get the maximum from the land they worked. Again we are fortunate that the worth of the tractor was recognised by so many enthusiasts, so today preserved examples are on show at steam and agricultural meetings throughout the country. While they may not get the acclaim that is given to traction engines, they deserve closer study, for without the tractor farming would still be in the dark ages.

9 · The Dairy and Dairy Farming

Today it is hard to realise that when American troops first arrived in England during World War II, they were warned against the dangers of drinking our milk. It was a fair reflection on the dairy industry as it then was. The keeping quality of much of the milk produced was sub-standard through lack of care over cleanliness from the cow to the bottle, and it was the exception rather than the rule to find a dairy herd tuberculin tested.

All that has now changed. Every milk-producing herd has to be tuberculin tested, and milk is subjected to regular checks for cleanliness. The body responsible for this change was not the farmers themselves, but the Milk Marketing Board. This came into being in 1933 as a result of chaotic over-supply, and while it naturally concerned itself with making dairy farming a financially viable venture for the farmers engaged in it, it also sought to improve both the quality of the product and the animals that produced it.

Through financial inducements to farmers to make their herds attested, and through compensation for the slaughter of tuberculous cows, the standard of the British dairy product has risen dramatically. Even more, the Board entered the technical field to improve the yield of the cows. The recording of the milk given by each individual cow enabled the farmer to eliminate the poor producers, and gave the information needed to enable a prosperous breeding programme to be undertaken.

The real breakthrough as far as the improvement in the dairy cattle is concerned has come through the Boards' artificial insemination programme. In a twenty-four year period since the scheme began, over twenty-five million cows were artificially inseminated with an increase of an average of two hundred gallons per cow.

In the past, bull selection was a difficult matter for the farmer, with many differing factors giving a confused picture when it came to assessing the worth of one bull compared with another. That is now a thing of the past. Extensive tests are run, checking on the yield of cows that come from different bulls with the result that the Milk Marketing Board, through their artificial insemination scheme, has built up dairy herds to a standard that would once have seemed impossible.

The predominant dairy herds in this country are made up of the British Friesian, cattle originally imported from the Dutch Province of Friesland. Friesians and other Dutch cattle came into this country during the latter part of the 19th century. While a large number found themselves at town dairies, some herds were established in the 1879's and again in the 1890's. However, few of them were kept as pure Friesians, and the breed as a whole made little progress until the establishment of the British Friesian Society in 1909. Its progress after this, however, was rapid and has continued that same expansion ever since. The strength of the breed was greatly increased with the importation of a number of specially selected cattle from South Africa and Canada, as well as its native Holland.

The standard colour of the Friesian, which is the largest of the dairy cattle, is black and white in distinct patches, although an occasional red and white one will be seen. In fact a separate herd book has been set up for this variety of the Friesian.

What makes the Friesian of such value to the dairy farmer is its great milking capacity. In most countries where it is to be found, the breed holds most of the records for quantity. In England, individual yields from 2,000 to over 3,000 gallons have been recorded, and one Friesian cow has given over 3,000 gallons in each of three successive lactations.

The predominance of the Friesian has been such that the Dairy Shorthorn, once the most popular breed no longer holds that title. However, it is not because of its milking qualities only, that the Friesian has attained such dominance. It is now recognised as a true dual purpose breed; cattle that can serve the dairy farmer, and also provide the lean meat with little waste to meet the needs of the butcher. So they serve both in dairy herds and for those whose need is for cattle that can be fattened for slaughter.

Among dairy cows however, the Ayrshire was one held in high esteem by Victorian farmers because of its milking qualities. On poorer land or at high elevation and in cold climates it was a valuable dairy animal, but today it amounts to only a small fraction of the dairy cows. The Jersey was another breed respected by the Victorians for the quality of its milk yields, and with a sandy coloured coat and deer-like head, they looked attractive as well. The Jersey, believed to be of French origin, has existed on the Island of Jersey for many centuries, and in fact no foreign blood has been permitted to be introduced since 1763. Under favourable conditions, the milk from the Jersey breed, with its high quality, is excellent for butter production. Another Channel Island breed is the Guernsey, the native cattle of Guernsey, Sark and Alderney. These are larger and less elegant than the Jersey, but the Islanders are equally proud of their cows, and strict import controls have preserved this breed.

Today the yield of the dairy cows goes to the Milk Marketing Board, who are responsible both for the fresh milk we use and that which goes for cheese and butter production.

Farmers of yesteryear had no transport system; each village was self sufficient, and there was no way to deal with perishable commodities like milk, the only alternative being to turn the surplus into butter and cheese. So on the dairy farm the most important building was the actual dairy, the province, not of the farmer, but of his wife and daughters. On their skill depended the quality of both butter and cheese. Cleanliness was an essential ingredient for success and the precautions taken against infection made a dairy a building apart from the rest of the farm. In the ultimate there would be separate rooms for milking and churning, and for cheese making and drying and storing cheeses. Even in those dairies where perhaps the churning and cheesemaking were carried out in the same room, attention would be given to hygiene and cleanliness. Flagstones in the floor were carefully joined and sealed to dispense with a source of dust, and the floor itself would slope in the direction of a channel that had been cut into it to carry water away into a drain outlet. The walls and ceiling were smoothly plastered and joined without any projection that could harbour dust. Double windows were usually fitted with the added protection of perforated zinc on the outside. An opening for the admission of air was built into the wall, but it was covered on the outside and had shutters inside that could seal it off when necessary. No foreign matter of any kind was admitted by a dairy maid intent on quality. In fact, such were the precautions for hygiene, cloths soaked in a solution of chloride of lime were hung across the dairy from corner to corner.

The output of the farm dairy varied according to local requirements. Butter could be made from cream or whole milk, ready for sale at the market, or salted in firkins for the family or dealer. Cheese would be made from fresh milk or skimmed milk and in any of the regional varieties. While cheese varied from region to region, in the main butter making followed a similar pattern throughout the country. After the cow was milked, the milk was strained through a sieve into a settling dish to remove any straw that could have fallen into the pail during milking. The settling dishes were made two to four inches in depth and about eighteen inches in diameter, a shape adopted to bring the maximum amount of milk into contact with the cooling air, thus causing the cream to rise. It was allowed to stand for anything up to twenty four hours before the cream skimmer transferred the cream to a storage vessel. The cream was

120

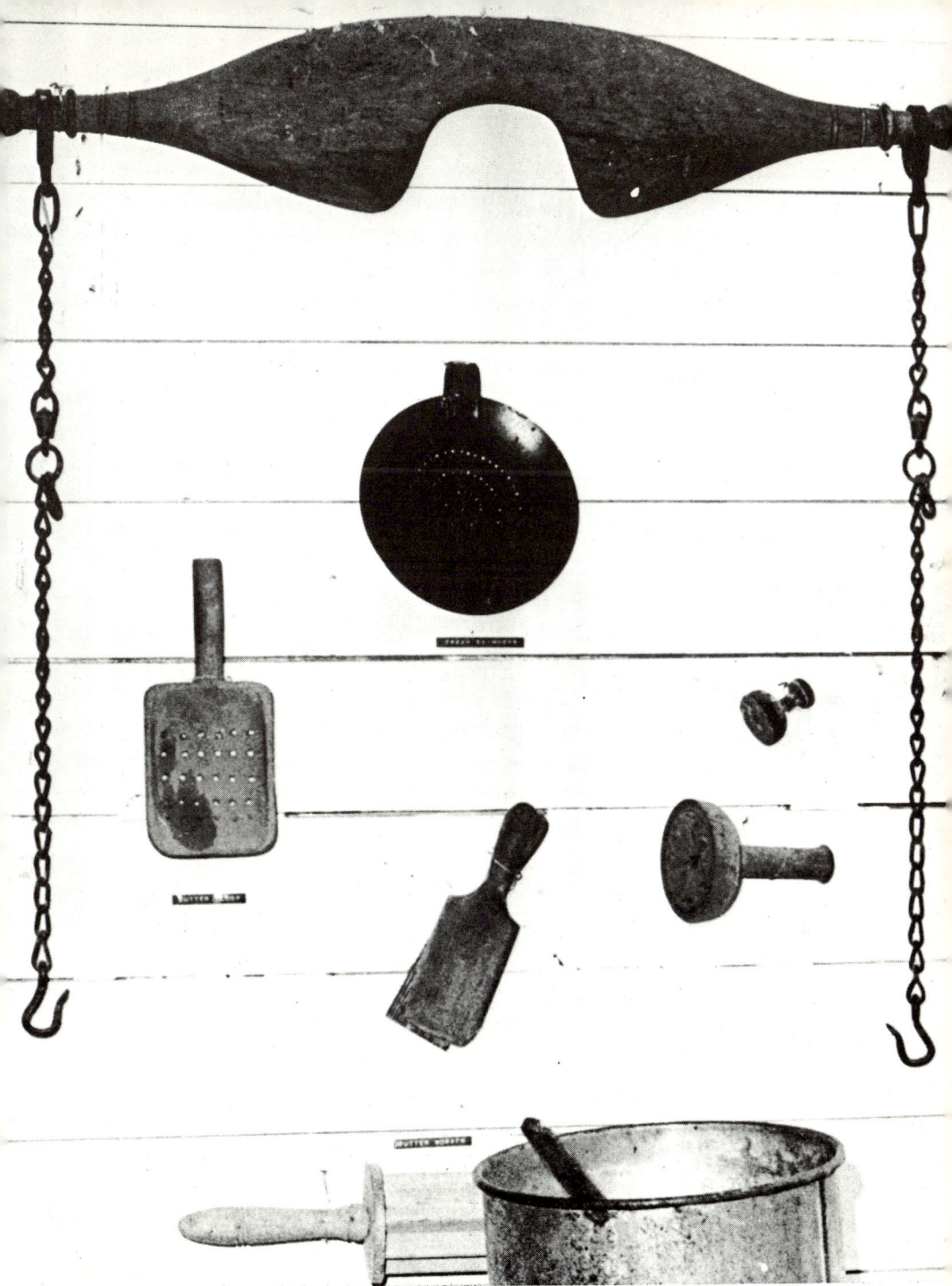

Old time dairy tools. At the top is the dairymaid's yoke, with hooks which were fitted to the buckets. Below is the cream skimmer, with butter pats and butter moulds.

Small butter churns used in the farm kitchen.

often transferred from vessel to vessel daily until sufficient quantity had been obtained for butter making to commence. This constant shifting and stirring prevented a skin forming on the top of the cream, which would have been injurious to the butter.

Butter churns were available in various shapes. The earliest, which replaced the old method of shaking cream in a glass jar to make butter, were the plunger churns. These were wider at the bottom than the top and had a close fitting lid with a hole to take the wooden handle of the plunger, which was worked laboriously up and down until butter formed.

The 'barrel churn' which came into being during the early part of the 19th century made the task considerably easier. Now a handle caused the barrel to revolve, and with the slat-work positioned inside the churn for the cream to flow through, the process was speeded up.

The churn was never the subject of haphazard approach. In summer, when the weather was really hot, the handle had to be turned slowly otherwise soft butter would result. Normally the churn was turned at about forty revolutions per minute, but no definite time could be given for completing the butter making and a single churning could last for many hours. Yet such was the skill of the dairymaids that many knew instinctively when the butter would be ready and were rarely wrong. In fact such was their skill with the churn that a good dairymaid knew exactly how much cold water to put into the churn in hot weather and how long to leave it to bring the churn to the right temperature for successful butter making.

In most regions the butter was washed when it came from the churn, but not in Gloucestershire. This was an area that had a taste for a sweet butter and common belief said that if the butter was left unwashed it retained its sweetness much longer.

All butter, however, when it came from the churn had to be worked to remove the last trace of buttermilk. Although wooden spades were made for this task, there were many women with firm ideas on this aspect of butter making. Butter made up by hand, they said, was more free from buttermilk, and of firmer texture than that which had been worked with the aid of spades. It was essential for the hands to be both clean and cool. First they would be washed in warm water and oatmeal, and then steeped in cold water before they were in a fit condition to touch the butter. To neglect to do this was to risk tainting the butter.

The crowning glory of the farm butter was the decorative motif proudly embossed on each pat. The image, often taken from heraldry, was carved on to a two inch diameter wooden stamp. Wooden 'hands' were longitudinal and transversal parallel lines were used to shape the final product into an attractive and eye catching shape.

Even in the 19th century there were scientific aids to help the dairymaid. Cylindrical glass cream gauges were available to show the cream content from the various cows. In operation they were simple indeed, merely glass cylinders graduated from 0 to 100 which showed the percentage of cream to rise after the gauge had been filled with milk and allowed to stand for twelve hours. Refrigeration, when it made its appearance from America in the 1850's, was another welcome aid for the dairy.

Today England is famous for its great cheeses, and it often comes as a surprise to realise that these originated in the farm dairy, variations being due to differences in the milks in many parts of the country. The reason can be found in the preference for one breed of cow in certain regions, and of course regional differences in the feedstuff and method of its production.

In cheese-making the milk was put into a large tub and warmed to a certain temperature. Rennet, a secretion from the calf's stomach, was added to turn the milk sour and cause curds to form, after which the liquid or whey was removed. Then the solid lump of curd was pressed and cut, and pressed again until all the whey had been squeezed out. If this part of the work was neglected and whey remained in the curd, it gave an inferior cheese that would cut wet and have a poor flavour.

The whey-less curd was then broken down and worked until it was of an even consistency, when salt was added. The whole curd was now put into a wooden mould of the size and shape of the final cheese. This cheese 'vat' was filled right to the top, when a metal lid was added. Then the vat was put into a cheese press for two or three days to rid it of any remaining whey. The final product only needed to be dried, but some reinforcement had to be tied round the sides to stop the cheese splitting before it was ready for market.

Some dairymaids believed that a cheese left on the floor to dry had a better flavour than a cheese in a rack, but wherever it was put the cheese had to be turned regularly, at least once a day, until the wrappings could be removed and any indentations filled in before the cheese received its final polish.

While English cheeses in general have an enviable reputation, in some regions there was one form of home made cheese that was not so appetising. In the period when unmarried farm workers would live in at the farm house, cheese was seen as an economical food for them, but this was made from milk that had already had the cream skimmed off it often, not just once. Spring was the time when batches of these cheeses were made, being allowed to mature through the summer months, and brought out only in the advent of cold weather. Then they were cut across, stood in front of a fire until soft enough to scrape and spread. Two thick slices of bread covered with this cheese, with slices of fat bacon in between, provided a cheap but nourishing breakfast for many farmhands.

The advent of railways and roads changed the face of dairy farming. Cheese factories came into being and the liquid milk industry was born. With the means of selling his milk to a wider public, part of the farmer's need to turn it into butter and cheese at the dairy was gone. In fact the demise of this side of the industry was so rapid that by the time of World War I butter was only made on farms where the geographical position led to transport problems, and fewer and fewer dairies were bothering with cheese.

Now that they could sell their milk, many dairy farmers increased the size of their herds, but few bothered to give any thought to the buildings that housed them, relying on the old traditional form, with rows of indi-

Butter churn used in the farm dairy.

The Plunger butter churns. The one on the right is quite rare, having been made of earthenware.

vidual stalls where the cows could be tied up in winter and milked all year round.

Gradually however, as milk began to be transported away from the farm, the health hazards with such a highly perishable commodity became more and more obvious. Although even during World War II these had not been eradicated completely, officialdom had already had a say in the design and construction of the cow house.

By 1885 local authorities were responsible for the issue of by-laws covering the layout of the cow house in regard to the general well being of the dairy herds. These and later regulations stipulated concrete instead of earthen floors, and roofs had ventilation outlets and roof lights. Dung channels were dug for effective drainage, and in new buildings both height and length were increased.

Once progress has started it has a habit of continuing, and this was true of the cow house. An inherent drawback, where the cows were contained in stalls during the period they could not be in the fields, was that it was expensive in terms of labour to both feed and collect the milk and manure.

Mechanical aids came into milking in the latter part of the 19th century, but these were no more than machines to transfer the milk from udder to bucket. Then in the 1920s came the first machine that was capable of carrying milk from the cow to the dairy through an overhead pipe. Although the carrying distance was limited, it was the basis of new ideas in dairy farming.

The idea was first conceived by J. Hosier, who was building a dairy farm on a thousand acres in Wiltshire and who needed the necessary milking buildings. In his case, the land was suitable for his herd to remain in the fields throughout the year, so he only required somewhere to milk them. To this end he built a lightweight milking parlour that could be transported to the field. Now instead of the cows having to be tied up at individual stalls, his system enabled each cow to merely file in to be milked, then file out again with another animal taking its place. The system was so successful that when Hosier decided to have permanent buildings he constructed what was the first of the modern milking parlours, where the cows came, one after the other, through the same stall to be milked. His design was suitable for any size of herd.

Progress in the milking parlour has seen developments lead to designs whereby the cows are arranged in herringbone fashion, with two banks of five, eight or even ten cows with back and udders pointing in a herring-

A hand-operated milk separator for the farm dairy.

bone pattern towards the operator who, with modern milking machines can handle fifty cows per hour.

While today we think of the milk parlour as the province of the rural scene, there were times when they were as common in our towns and cities, where the need for milk was the same as in the country, but where through lack of transport facilities, its transfer from one place to another was impossible.

The first records of such cow houses date back to the 17th century when London's population was expanding at such a rate that it was outgrowing its milk supplies from the surrounding countryside. In one respect this was the first example of factory farming, for newly calved cows were bought to be milked through their lactation and then fattened up for sale. Large numbers of cows were kept in cities like London and, by the middle of the 19th century the cow population exceeded twenty thousand. The coming of a transportation system enabling the dairy farmer to supply the needs of urban dwellers spelt the end of the city cow house, but in London some persisted up to 1930. Today they are but another example of the past that has gone for all time.

No matter whether the cows were in town or country, they had to be milked. In the early days the milker sat on a three legged stool and milked by hand. It was effective but slow, and so thoughts turned to the provision of a mechanical aid.

Various attempts were made at producing a milking machine but none was successful until an American called Colvin came up with a hand operated vacuum milker in 1862. The idea failed because although the vacuum principle was right, there was nothing on the machine to imitate the pulsating action of the calf suckling.

William Murchland brought out his milking machine, again on the vacuum principle, in 1889. Designed for permanent installation, it utilised a vacuum created by a column of water and continuous suction. The machine proved that it could do the job, but continuous suction proved not only painful to the cow being milked, but also caused damage to the udder.

The man who found the answer and paved the way for the modern milking machine was a Scotsman, Dr Alexander Shields, who introduced his pulsating system in 1895. Unlike other vacuum milkers, Dr Shields added a device that caused a regular brief break in the suction, in the same way as a calf would behave during natural feeding. It was expen-

Butter basket, butter pats and embossed moulds.

sive, but it was the answer to the problem of finding a milking machine
that could be used without causing damage or suffering to the cow.

The modern milking machine has an action that can be broken into two
parts; sucking and squeezing, accomplished by the weight of the equipment hanging from the teats. A vacuum pump is driven either by an electric or small internal combustion engine in the modern dairy installation.
Then there is a sanitary trap which prevents water or dirt being sucked
into the pump. In most dairy systems the vacuum pipe runs along the wall
with stall taps being provided to take the connection from the actual milking bucket which has the pulsator on top. This is the device which draws
air from, and lets it into, the teat cup assembly which is attached to the
cow, giving an action identical to that of the calf suckling.

10 · Livestock

To the early farmer one of the most attractive animals must have been the humble pig. For in the age when feedstuffs were always in short supply, the pig was almost self-supporting, spending a large part of the year feeding on the beech nuts, acorns and grubs to be found in abundance in forests and woodlands.

In fact, in village life the pig continued to play an important part until comparatively recent times. There would be few cottages in the rural communities of England that did not keep a pig. In many areas it was a financial necessity, something that tradespeople, from the tailor to the bootmaker would give credit on.

Although they may not have realised it, those cottagers who followed tradition and housed their pigs in the seemingly unhealthy low stuffy pig sties that had survived in village life from generation to generation, gave a lesson to the first designers who tried to introduce clean, well-lit buildings for the use of the pig farmer.

As architectural designs they were an improvement on what had gone before, but what was not taken into account were the requirements of the animal they were going to house. There is no record of who first realised that the survival rate among young pigs was consistently higher if they were reared in the old primitive conditions rather than in the new type of offering. The pig sty attached to so many country cottages had evolved because early farmers watched their animals and knew that in its first period of life it needed, above anything else, warmth. In the wild state, as our farming ancestors knew, the mother made a nest for her offspring during the first weeks of life, a factor copied when man began to make a sty for his pig. The first builders who sought to modernise the pig's abode forgot this until they discovered that the pigs belonging to the cottagers were more healthy than those of the farmer.

Above all, the pig shows, as possibly no other animal can, the fetish our ancestors had for seeking the ultimate in weight and size from their animals. To read some of the claims for pigs, in particular during the mid-19th century, is to realise just what size was achieved.

"If a pig can walk two hundred yards he is not fat," wrote one agricultural expert during the Victorian period, and the Yorkshire, or Large White was one breed that proved the truth of his words. Two prize-winning

Yorkshire pigs, weighed in 1856 when they were three years old, were each over 11½ cwts, and one was 7'2" long, from the ham to the end of her nose, with a girth behind the shoulders of 7'8". Neither was this an isolated example. A claim made three years later was for twelve Yorkshire sows with a combined weight of just under six tons. Pigs such as this however, came in the period before science was brought into breeding. Prior to 1884 breeding was a haphazard business, but that year, in an attempt to adopt some uniformity, the National Pig Breeders Association was formed.

Farmers began to look outwards, bringing in breeding stock from China and the Mediterranean countries, to improve our domestic pigs, decreasing weight and gaining an improved carcase. Through cross breeding and selection, the 'Large Whites', 'British Saddleback', 'Berkshire', and 'Gloucester Old Spot', appeared on the farming scene, together with others that soon passed into oblivion.

However, the modern British farm pig owes its origins to Denmark. For it was the Danes who imported the British Large White, and breeding with it produced the 'Land Race' of pigs that, in the story of progression, we brought into Britain in the 1950s.

The modern pigs today are hybrids bred for profit. Crossing the 'Land Race' with such as the 'Large White', pigs have been produced with a narrow front end, the cheaper cuts off the carcase, but with big long backs and back ends, the most expensive meat. In fact the modern pig is reared to give the maximum weight return against food consumed, with that weight being in the parts which will fetch the highest price when sold in the butcher's shop.

Sheep are another animal now reared and bred to take into account public preference. The aim here is a small compact lean sheep and this has been achieved with the hybrid breeding of British and continental stock.

The position today is very different from that which existed in the middle of the 19th century, when the important sheep rearing areas were to be found in the arable corn belt of East Anglia and the southern parts of the country. There they were valued for the soil fertility, and the way they helped prepare the land ready for corn by consuming roots and other crops, as well as treading the ground.

By the end of the 19th century the arable picture was changing. Corn prices fell, and the public began to demand smaller, leaner joints of a type being imported from New Zealand. Up to then it was breeds like the 'Lincoln' and the 'Cotswold', suitable for the arable farmer but which mature into fat, heavy carcasses which had been popular. These breeds supplied the wrong sort of joint so even then there was a trend towards a smaller sheep. It was sheep such as this smaller stock which saw the shift in emphasis that has carried on into modern farming, for they marked the colonization of the hill country with sheep, so that hill farmers began to devote most of their efforts to sheep farming, supplying lowland farmers both with ewe lambs and older ewes capable of a few years breeding.

Sheep, of course, provide a return other than their carcase, and to the sheep farmer sheep shearing is as important as the harvest to the arable farmer. Before shearing could be done the sheep were washed to remove dirt and impurities from the fleece. In some cases this meant nothing more than forcing the sheep to swim across some natural river or stream, but the more usual method was to construct washing pools in which men stood, washing each sheep as it was forced into the water. Washing was usually done early on a summer morning so the sheep would have the rest of the day to dry out. Usually a further week went by before the shearing operation was ready to begin. This allowed time for some of the natural oils, removed by washing, to return.

Today, washing has no part in the shearing of sheep, but another essential for good health, the sheep dip, has. Dipping was the process by which the sheep were taken through a prepared bath of insecticides with the object of killing external parasites and preventing attacks by blowfly. In some parts of the country there were regulations making it necessary to use a winter dip, particularly against sheep scab. Dipping, in fact, almost fell into disuse, but the return of certain diseases has seen it reintroduced to safeguard the flocks.

In the early days shearing was done with a tool that looked like a pair of large scissors, blunted at the ends to avoid injury to the sheep, and able to be used in one hand. They were followed by shears with a sprung arc at the top to make action easier. Even with these primitive tools, experts were able to sheer fifty or more sheep in a day. Their drawback was the fact that they had to be guided over the sheep's body and at the same time worked in a clipping action.

So the arrival of a cutter that reciprocated between the combs and forks of a shear head, being driven first by hand through suitable gearing and later by electric motors, was a much needed advance. This was the basis of the shearing cutters which are used today, powered by electricity.

Although cattle have always formed part of the farming scene, their numbers in the early days, were decided by the amount of hay available to feed them through the winter. For a farmer in that age had no alternative but to slaughter the cattle he could not feed, and salt down their meat for himself and his family. It was, however, a system which worked against those trying to earn a living from the land in the days when cattle such as the oxen were the main beasts used in agriculture. The number of such animals which the farmer could bring safely through the winter determined the amount of power that would be available to work the land, and the amount of manure that would be available to enrich it. As this was the controlling factor regarding the next corn harvest, food production, in those days was very precarious indeed.

The gradual advent of new crops, especially the turnip which was so essential to the Norfolk four-course rotation, and other crops such as the mangel that made its appearance in the fields about 1819, meant that the need for slaughter had gone and the farmer could ensure that his best beasts were kept for breeding purposes.

Many of the old breeds have long since disappeared, and those that have survived owe much to the work of early pioneers, dedicated to improving British cattle stock. The brothers, Charles and Robert Colling, who farmed near Darlington, gave their selective breeding care to the 'Shorthorn' cattle.

The important 'Shorthorn', while its ancestry is not known for certain, appears to come from the old black Celtic, the red Anglo-Saxon and imported Dutch stock. While some attempts were made to improve the stock before 1780, it was the efforts of these two men, who began their selective breeding in that year, which gave us the modern Shorthorn. Their noted herd began with one selected bull and four cows, and they carefully built it up, eradicating the usual faults through interbreeding until, at Charles Collings' herd sale in 1810, his bull 'Comet' was the sensation of the sale, fetching one thousand guineas.

Other breeders followed the example set by the brothers, among whom was the Booth family of Yorkshire, famous through three generations as Shorthorn improvers. In fact it was John Booth, who in 1818, brought about the evolution of the Shorthorn as both a beef and dairy cattle.

As a breed, Hereford Cattle were held in high esteem from as early as the 17th century, both through the work ability and meat quality. It was the inherent high standard that drew men like Benjamin Tompkins, who began breeding experiments in 1766, to improve on this already fine line of cattle. The success that he and others achieved gave the Hereford the highest combination of early maturing and fattening qualities, with a

134

Sheep shearing.

strong constitution. It will fatten easily on good grass alone, but abroad it has proved its ability to thrive on the poorer dry grass where other breeds cannot exist at all. While the quality of the meat from the Hereford brought favourable comments from our early farmers, for true superlative quality the beef prize must go to the famed Aberdeen Angus. This breed, which originated in the old local breeds of Angus and Aberdeenshire, can date their family tree back to the middle of the 16th century, when their black hornless characteristics were already established.

Breeders naturally turned to the Aberdeen Angus to try and improve this excellent animal even further, and of these possibly the greatest was William M'Combie of Aberdeenshire. Not only did he succeed in improving the qualities but sang its praises so loud and far that he established the universal reputation that it now holds. While the breed is not the largest of our native cattle, the Shorthorn, Sussex and Hereford all being bigger, the Angus is compact and heavy for its size, and in open competition at the various fat stock shows over the years it is the breed which has won more championships than all other breeds put together.

There are, of course, other types of British cattle. The Galloway, an old breed which two hundred years ago occupied, to the near exclusion of other cattle, most of south-west Scotland. This is one of the hardiest of our cattle, capable of withstanding the worst of weathers, although it is brought down from the hills to the lowlands in the winter. The Belted Galloway with its characteristic white belt round the middle of its body is a distinct strain of the breed, and the only belted cattle left in Britain. The Sheeted Somerset, another belted type, saw its last herd slaughtered in 1934 when it failed to pass a tuberculin test in the period when efforts were being made to improve the quality of British milk.

Among the remaining breeds, the most notable are: the Dexter, the smallest cattle in Britain; the Lincoln Red, a strain of the Shorthorn found mainly in its native county; the Charolais, from France; the Chianinina from Italy; and the Swiss or German Simmental. All these breeds number among those to be found on modern farms.

The horse has been largely lost to the farm, the victim first of steam, then of the internal combustion engine, but recently there has been a surge of interest in the old breeds that once looked to be in danger of dying out. In the same way as steam has been saved by the preservationist, so many farmers are turning to the old draught horse. Interest has escalated to such an extent that old skilled horsemen are offering tuition in the arts of horse ploughing, and their pupils and others are there to show the 'period of the horse' at Agricultural Shows, at Ploughing Matches and other events where the old traditional methods can once again be seen.

Possibly the most famed of all working horses is the Shire, regarded as the direct descendant of the old English war horse, the horses that carried England's monarchs into battle. In fact the Kings, from John to Henry VIII put in considerable effort to increase the size and quality of their horses.

The present day Shire is possibly the heaviest of all horses, and Victorian breeders were advised to concentrate on making their stocks as large as possible. Demand then existed in our towns and cities for a single horse that could haul large and heavy loads on its own. It was the period when urban areas were overcrowded and hauliers were crying out for a horse capable of working on its own, rather than having to use a team with the attendant difficulties caused by their manoeuvrability.

Pulling power was essential with all draught horses and in Suffolk, the home of another famous breed, the Suffolk Punch, the drawing match

Two champions of the Shire horse.

was an occasion when all the village would turn out to see these horses in competition with each other.

The test was made on a loaded farm waggon with its wheels partially sunk into the ground and with wooden blocks fitted in front of them. It was a test of the most notable quality of the Punch, and proved its ability to pull a dead weight. Such was their worth in this respect that the horse would drop on to its knees to apply the maximum exertion to the load.

The third native British draught horse is the Clydesdale, which had its origins in the first half of the 18th century through the interbreeding of Scottish and English stock, with some of the latter coming from a Flemish breed. In fact one of the famous Clydesdale breeders, John Paterson of Lochlyoch, in about 1715 introduced a black Flemish stallion from England which was believed to have started the line. They soon became the dominant breed in Scotland, possessing endurance, strength and hardiness. The earliest of the line to which pedigrees can be traced today is a horse foaled about 1810 and believed to come from the original Lochlyoch stud.

Livestock of course, was prone to sickness and ill health and the farmer often had to be his own veterinary surgeon, applying medicines as well as docking horses' tails and branding his stock so they would be easily recognised.

The appliance for administering medicine was the 'balling gun'. The drugs to be given were made up in the form of pellets or balls, and 'shot' into the animal by the 'balling gun' a hollow wooden tube that had a plunger handle attached to it. The mouth would be held up with a 'mouth clamp' and the'gun' used to fire the medicine home.

Branding would be done with a tool made by a local blacksmith, and usually consisted of the farmer's own initials forged in iron to withstand the heat necessary to apply the mark. Today branding irons have fallen into disuse as a specially prepared indelible marking material has become available.

Then there was the docking of the horse, the amputation of the last six joints of the tail to prevent its long point being injured when the horse worked in the cart or waggon. The implement for the task was the 'docking iron'. First the tail was prepared by cutting away the hair at the sixth joint, which was then laid in the circular notch of the docking iron, before the cutting plate was forced down, severing the unwanted part of the tail. Docking now is illegal, yet another aspect of the cruel side of farming that has been prohibited.

II · The Village

Today the village as it used to be, has been lost for all time. The close knit community, almost self-sufficient, has been replaced by housing estates for those who turn to the town or city for their daily living, yet seek the tranquillity of the countryside when work is over. Even those born in the village more often than not have to seek employment outside.

True, the general store and the odd butcher's shop have survived to supply some of the needs of the inhabitants, but the advent of the motor car and the growth of the super- and hyper-markets has meant the end for most of these old country retailers.

Yet how different the position was even at the time of that great tractor trial held in 1920 near Lincoln. The village then was a unit in every sense of the word. To quote the small Lincolnshire village of Ropsley, which in those days had a population of 519, and whose three public houses would have vibrated with the comments of those fortunate enough to have attended the trials, is to show how life has changed. In addition to the three 'locals', the village supported two bakers, a tailor, three butchers, a plumber, boot-maker, saddler, three grocers, an insurance agent, two joiners, a miller, mason, and blacksmith as well as two general shop-keepers and a threshing contractor.

Today progress has prohibited their trade. With the exception of the two public houses, a general store and butcher's shop, Ropsley people turn to the shops of nearby Grantham for their needs.

Even in the year when the tractors were convincing the farming community of their potential, rural Lincolnshire was dominated by the horse and steam, for there were more than 50,000 horses employed in agriculture in general, and a total of 144 agricultural steam machine owners. Some of these would be small, doing their own work and possibly that of a few nearby farmers, while others worked on a massive scale and brought some of the few outsiders into the village in the form of 'contract men'.

These men who worked for steam contractors like the Sleaford firm of Ward and Dale, were a class apart. The firm itself was one of the leading Steam Cultivating Companies in existence in the 1880's, becoming a limited company in 1910 with assets including ploughing engines valued at over £15,000. They were known not only in Lincolnshire but in Nottinghamshire, Rutland, Leicestershire and even Northamptonshire, where the pairs of massive Fowler engines used to work during the plough-ing and cultivating seasons. In fact, during 1914, nearly 65,000 acres of land were cultivated by Ward and Dale.

Each pair of engines needed a crew of five; two drivers, a steersman for the plough, a foreman in overall charge and a boy, usually straight from school who acted as cook and general labourer. The work, during the April to December season, was long and hard. The farmer hiring the crew rarely allowed any slacking. The equipment had to be kept working even through meal times. When the boy cook signalled that the mixture of boiled bacon and potatoes was ready, each man in turn would 'wine and dine' using the foreman as relief crewman.

Ploughing or cultivating started at first light and went on until it was too dark to continue. Then the tired crews would collapse in their bunks, (more often than not fully clothed) in the living van that went with them from job to job. Some weekends saw the men returning home late on Sat-urday, but one at least had to be back with the engines to light the fires by 4 a.m. on Monday so that steam would be up and work could commence as soon as the others arrived.

Life for these men was hard and wages poor. The only concession they received was a bonus depending on the amount of acreage covered, and this was withheld by the firm until the season was finished. The more thrifty could survive financially on this until the following year.

Adequate cultivation is essential for success in agriculture, and while the Steam Contractors may have been itinerant workers as far as the vil-lagers were concerned, they played their part in the life of the rural com-munity.

The miller, of course, depended on a bountiful grain harvest. In contrast with the men who came and then left when their cultivations had been completed, the miller was a man of convention, handing down his job from father to son, so that generations worked the same mill in the time honoured ways.

Windmills in England fall into three types. The oldest mills, those in existence at the time of the Doomsday Book, were all post-mills in which the entire mill could be swung round to face the wind. It was only in the 18th century that the tower mill came into being. This consisted of a round tower of brick or possibly local stone which housed the milling machinery. Only the cap, which contained the axle for the sails, turned in the wind, going round in a similar manner to a turntable. The third design was wooden with an octagonal tower, commonly known as the 'smock

A charming reminder of a lost age — a well bucket and three-pronged drag.

mill' because of its resemblance to the smocks worn by countrymen in
those times.

A necessity for efficient milling was to keep the sails facing into the
wind, and until the middle of the 18th century this was done by manual
labour working the tail pole at the back. Later, however, a wheel of vanes
applying the fan tail system was installed at the back of the mill to hold the
sails in the required direction. The windmill is representative of the very

Tower mill, Heckinton, Lincolnshire — the only one remaining with eight sails.

best in rural craftsmanship, for the mass of axles, wheels and spindles which carried the power from the sails to the grinding stones themselves were all hand made and fitted together with a perfection that today we can only marvel at.

Although the diameter of the stones was no more than about four feet, their weight could be as much as one ton. Sources of good material were few, with the best being found on the Continent. The weight alone made transport a slow and costly business, and many millers were forced into using a hard sandstone quarried in the Derbyshire Peak district and known as Derbyshire stone.

The best of all stones however was without doubt French Burr. These were pieced together from pieces of a variety of quartz quarried near Paris. Found in small deposits only, the material had to be shaped and matched before being joined together with cement and then bound with iron hoops.

Millstones were arranged in pairs. The lower, or bed stone, was stationary, while the upper 'runner stone' rotated face down, the separation distance being under the miller's control, a vital factor if the best milling was to be achieved. The stones had to be 'dressed'; that is, an arrangement of grooves cut into them. Some millers, skilled in every aspect of their trade, would do the work themselves. Others employed special dressers. This was hard demanding work, with the 'dressers' working from six in the morning to six at night, and until five on a Saturday. If the demand was there from the millers, then they would toil away until the work was finished, whatever the time, and for this they were paid the princely sum of four pence per hour overtime rate.

The miller's life was not always a safe one. For if a violent storm struck it could be very dangerous indeed. The mill, when nature turned against it, could rock from side to side, sacks of meal would be thrown about inside, and if the miller tried to apply a brake to the racing wind shaft there was the risk of setting the whole mill alight through the friction created.

Many village mills, such as the one at Ropsley, have disappeared completely, remembered only in names like Mill Place, Mill Hill and Mill Drove. Others, both wind and water, have survived, even in the same family that has served them for generations, to show yet another example of a lost way of life.

The water mill at Ollerton in Nottinghamshire is one example of a mill that was handed down, from father to son, and today is still in the same

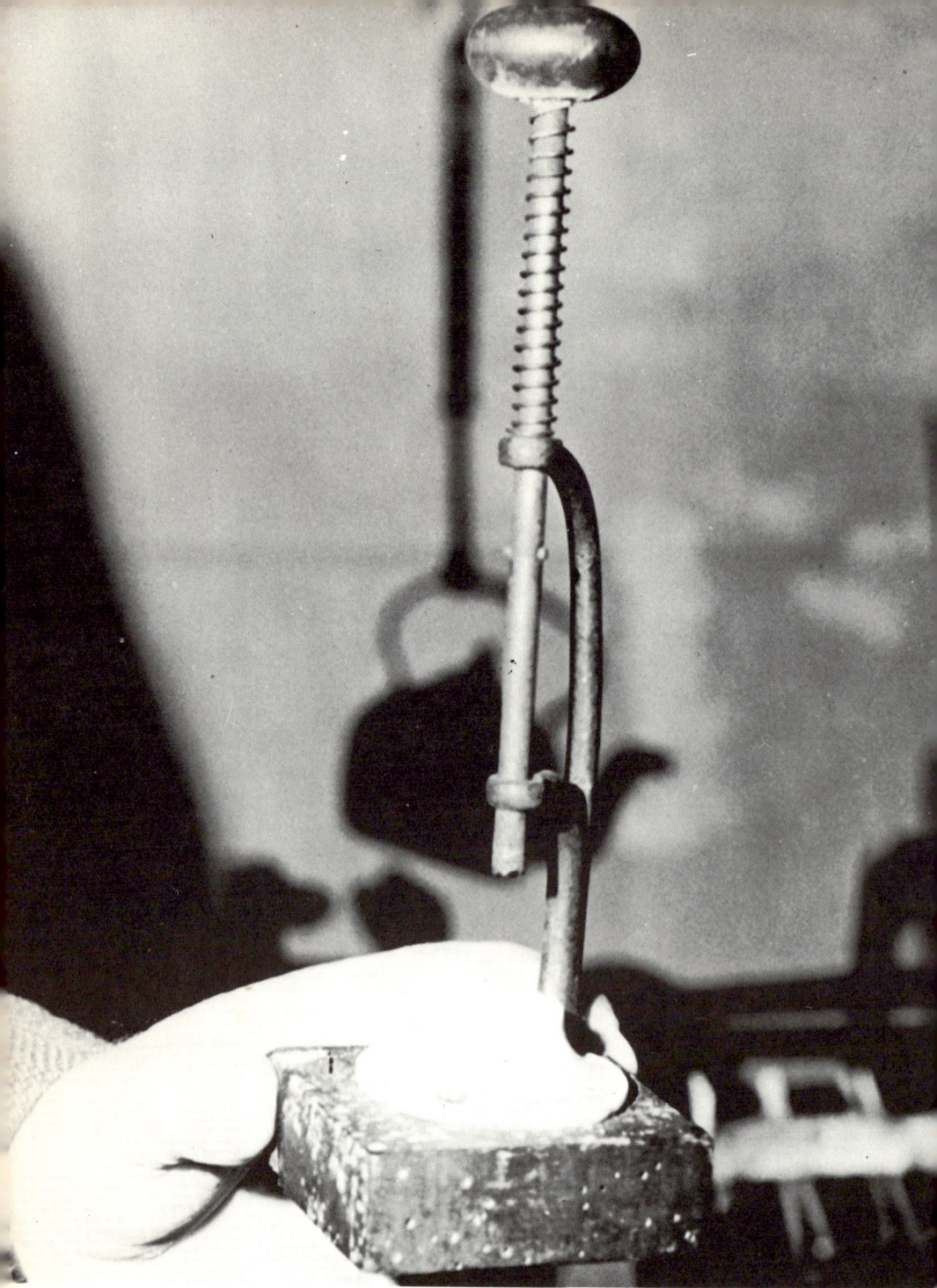

Centre-loaded cherry stoner.

A simple apple peeler.

family and in working order, although the farming evolution has seen it fade from the active scene.

Different indeed from those hectic days during the Second World War when the water wheel was turning continually twenty four hours a day, seven days a week, with the only concession to rest being a break of six hours during Sunday evening. The output then was 1,000 bags of flour or animal food a week. Yet such is the perfection of work of those skilled craftsmen of yesteryear that the mill is still capable of resuming an active life. In fact so precise was their work that a narrow groove had to be cut into the brickwork of the mill wall to give the necessary clearance for the maindrive wheel, and today that same wheel follows exactly the same prescribed circle.

Pair of ceremonial cannon used for passing news to out-lying parts of an estate.

Yet while the village in the 19th century was a self contained unit with tradesmen to serve all needs and requirements, it could never be really independent of the outside world. The saddler had his requirements that could not be met from within; in turn the tailor would need cloth and threads. In fact whatever the business there were certain goods that only an outside supplier could provide. This was where the village carrier came into his own as one of the most important among all tradespeople. It was the carrier who brought in not only commodities from the outside world, but relatives making a long-promised visit to someone in the village. The carrier was often the only man who could take some anxious villager on a trip into town.

To read through the records of such men, particularly during the mid-

19th century period, is to rekindle a lost way of life. For this was the age of the 'sugar loaf', those tall cones of sugar that the local grocer used to buy and chip into lumps with his sugar cutters before selling it to the villagers. In those days treacle was brought in by the carrier in barrels. In fact, for the general store, almost everything from soda to coal came in bulk. Nor was the life of the carrier an easy one. He was expected to make his journeys regardless of the weather. Snow, ice or fog made no difference to a countryman who often found himself working an 18 hour day, being on the road as early as four in the morning, winter and summer.

An early start such as this would mean the carrier loading up the night before and checking that his shopping list was up to date. Not that hard work upset his sense of humour. For one noted Lincolnshire carrier, when he had passengers going into market or visiting relatives in a nearby town or village, would charge first, second and third class fares. The rea-

The farrier's tools. Bottom left shows two mouth gags for horses, and above, the 'gun' for administering pills.

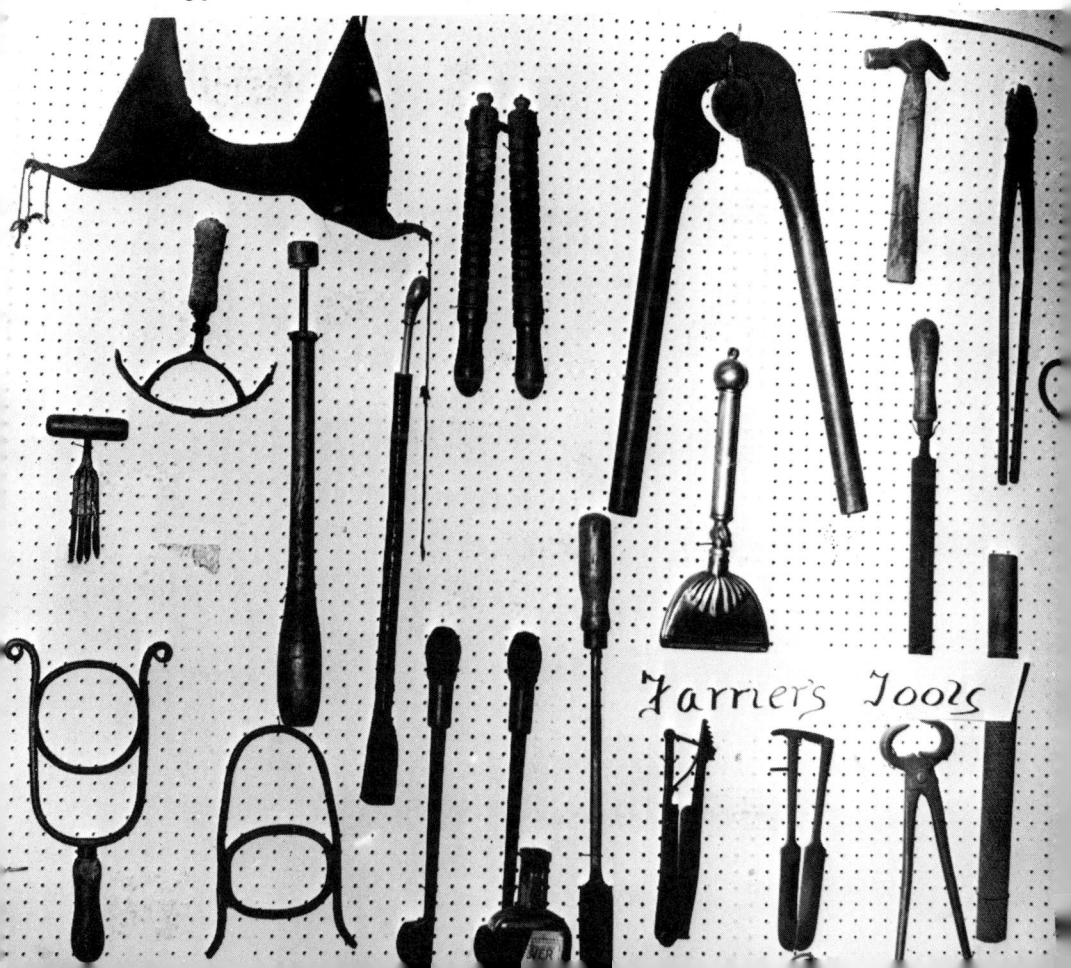

son was revealed at the first steep hill. First class sat tight, second class got out and walked, while those who travelled third class found themselves pushing!

The carriers were characters on their own, men respected and welcomed in both village and town where they took their orders to local shopkeepers, often receiving a small commission for the business they brought. More than just serving town and village, they were the only link with the outside world for weeks on end for those in hamlets and isolated farmsteads where it was the carrier who brought both mail and news. He did much more than supply the trading needs of the village. It was the carrier who would take to town the farm butter, the first of the new potatoes, eggs, poultry, in fact everything that was made on the farm for sale outside. Their rewards were not high. In the mid-19th century the cost of parcels was from 2d upwards, and considering that most needed two horses for their work, there were few chances of great financial gain. The carrier

Victorian clockwork bird scarer.

A man trap of the kind used to trap poachers.

however, like so many other countrymen, saw their work, not for gain but as a way of life that had been in their family for generations.

In fact hard work was synonymous with village life in the period when it was a self-contained community drawing a living from the land. Even children's education was secondary to the needs of the farmer. In fact education in many villages was often directed at only equipping the child to follow in the servant farm worker class. Many teachers saw a higher education as making the child discontented with his place in life. It was an outlook that saw young people leaving school able to read and write, but with little other academic qualifications.

Small wonder that when harvest time came round the school teacher would co-operate with the farmers and close the school so that the pupils

Examples of the various sizes of mousetraps in use during the Victorian period.

would be available when needed on the land. When the harvest was finished, then the children, and in particular the girls would be needed for gleaning — collecting the grain that still remained on the harvest field.

Gleaning played an essential part in the daily lives of the villagers. The grain collected was ground down, often being added to the corn that the farm worker grew himself, to provide flour for bread making. In our old rural communities, gleaning was far from being a haphazard business. The women and girls were only allowed in the fields when the harvest work had finally been completed, often by farm workers raking the ground to recover as much as possible of the grain that had been dropped. Even when the field was clear, they had to make an organised entry, signalled in some villages by the ringing of the church bell at 8 a.m., and gleaning continued until the ringing of the evening bell brought proceeds to an end.

Another activity that took the women and children into the fields was stone picking, done both to clear a field of stones and provide material for road repairs. It was a back-breaking business usually detested by all who took part, yet although the pickers received less than a penny for each bucket they collected, the state of the average farm worker's finances made stone collecting an essential task in the battle for survival.

Bird scaring at seed sowing time was another job where the village boys could earn some extra pence, but even this seemingly simple task demanded endurance. Long hours were spent in the fields either shouting, swinging hand bells and rattles or using the old clapper scarer. Clappers made by the boys themselves consisted of a central piece of wood with separate pieces on either side. These were loosely joined together and made the desired noise when the users walked the fields, clappers swinging in each hand.

While life as it was in rural England has been lost for ever, there are surviving examples that show both the simplicity and ingenuity of those who lived through the 19th century in particular.

The clapper scarer may have been primitive in design and use, but the same cannot be said for other types of bird scarer. These would often be cleverly made using a clockwork motor that controlled a metal arm, falling at regular intervals on a revolving drum containing side fire blank cartridges. The cover itself would be tastefully decorated depicting a bird in flight. The only trouble with a bird scarer such as this was that it needed frequent attention replacing the cartridges that had been used.

To the farmer, birds have always been a source of nuisance and worry, devouring both seeds and crops, but game birds, like the partridge and pheasant, were looked upon in a very different light. The methods used to prevent those who sought unlawfully to augment their diet with a delicacy such as these, were cruel and harsh. Man-traps capable of severing a human leg were cunningly placed on the paths that poachers would be likely to use. The cruelty was not limited to men, but extended to the dogs they brought with them. One harsh method of dealing with dogs was made in the form of a metal hoop with a spike protruding both upwards and downwards at the top. It was positioned in a hedge or undergrowth, tempting the dog to wriggle through with the inevitable injury from the downward pointing spike. If the dog attempted to jump, it was so arranged that the upper spike would find contact with the underpart of the dog.

Mousetraps of course were as necessary on the farm then as now, but in

A dog spike used to protect game. This would be positioned in a hedge and when any dog attempted to go under or over, it would receive serious injuries.

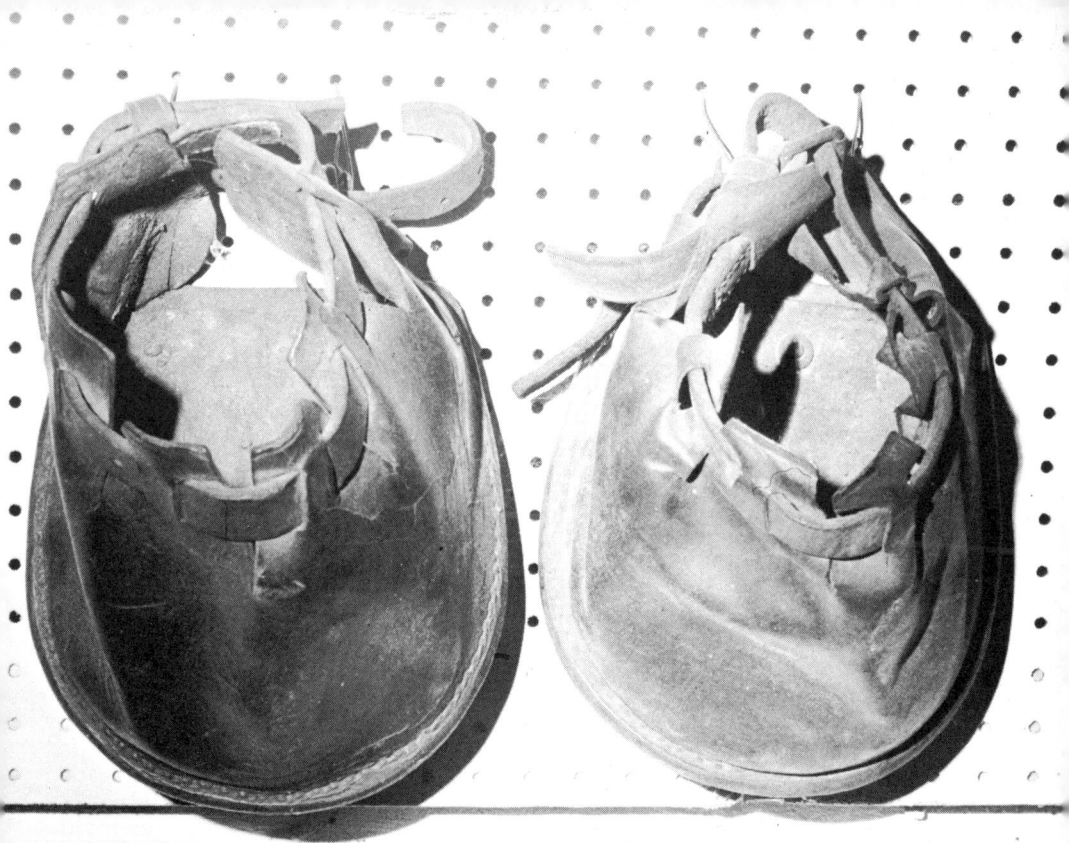

Horse boots used when horses were working on newly-laid turf.

contrast to the utility approach that we use today, the Victorian trap was a tribute to craftsmen of old. Made of wood, they were so designed that when the mouse entered, attracted by the bait, a heavy wooden block descended with a speed that made escape impossible.

Wood was also ingeniously used in an aid designed to help the farmer's wife when it came to lemon squeezing. Here two blocks were hinged together, one being recessed to receive half the lemon, with holes for the juice to flow out. The other block was exactly opposite, having a protruding dome that fitted firmly into the opposite recess. In use, the fruit was inserted and pressure applied, pressure that was guaranteed to extract the maximum juice from the fruit. Equally ingenious was the Victorian approach to the problem of removing the stone from a cherry. Here a pot-cup received the fruit and the extraction process was accomplished by

153

pressing down a spring plunger. Even the humble apple was not beyond the concept of the labour saving devices that were then available. The apple peeler that would be found in some of the more wealthy farmhouses was really a masterpiece in iron work. The apple was inserted, a frame closed round it, and as the handle was turned so a blade removed the skin evenly and cleanly.

Another interesting relic of village life, was a product of the local saddler, known as 'horse boots'. Made of leather they were strapped over the hoof, protecting newly laid turf and similar surfaces from damage by the conventional shoe. There are those who see a more sinister use for them and speak of the times when the coaches of the body snatchers travelled at night along our country roads, passing almost noiselessly because their horses were fitted with similar leather shoes.

Rural dwellers in the Victorian period of agriculture, and indeed even beyond, would have had no piped water supply. They relied on the well, and to this end they would turn to the village blacksmith to make a strong iron bucket, with the necessary hook on the handle for the rope. There were those occasions when the rope broke and a three spiked recovery hook had to be thrown down as grappling began. Today these are almost forgotten aspects of the past. But like so many other sides of rural life, enthusiasts, recognising the value of our rural heritage have searched for and preserved living reminders of what was once normal day-to-day life in the countryside.

Not that the purpose of everything they have found has been immediately obvious. For many rural areas had their own customs and uses. In fact a first glance at a pair of ceremonial cannon to be found in one private collection may defy belief as to their purpose, for these were never intended for the battle front or even the army. Nothing more than pieces of iron tube secured to blocks of wood, their importance in rural life came when the Squire of a large estate wanted to pass on some particularly important news, maybe the birth of a son and heir, a victory for the British in battle, or even the death of a member of the family. To have informed everyone individually, including those living or working on the extremes of the estate would have been a lengthy business, so these humble cannon were fired and simultaneously their message was heard by all.

*   *   *   *

That age has gone for ever, but reminders are all around us, in museums devoted to recapturing the countryside as it was, and in exhibitions held by those dedicated to the task of preserving farming equipment and everything that went with it. To those people we owe a debt of gratitude, for without them our own lives would be the poorer.

Index